The
ORCHARD
BOOK

Plan, Plant and Maintain
Fruit from Garden to Field

WADE MUGGLETON

Permanent Publications

Published by
Permanent Publications
Hyden House Ltd
The Sustainability Centre
East Meon
Hampshire
GU32 1HR
United Kingdom
Tel: 01730 776 582
Email: enquiries@permaculture.co.uk
Web: www.permanentpublications.co.uk

Distributed in North America by
Chelsea Green Publishing Company, PO Box 428, White River Junction, VT 05001, USA
www.chelseagreen.com

© 2021 Wade Muggleton
The right of Wade Muggleton to be identified as the author of this work has been asserted by
him in accordance with the Copyrights, Designs and Patents Act 1998

Cover illustration by Sarah Robinson, www.sarahrobinsondesigns.co.uk

Designed by Two Plus George Limited, info@twoplusgeorge.co.uk

Printed in the UK by Bell & Bain, Thornliebank, Glasgow

All paper from FSC certified mixed sources

The Forest Stewardship Council (FSC) is a non-profit international
organisation established to promote the responsible management of
the world's forests. Products carrying the FSC label are independently
certified to assure consumers that they come from forests that are
managed to meet the social, economic and ecological needs of
present and future generations.

British Library Cataloguing-in-Publication Data
A catalogue record for this book is available from the British Library

ISBN 978 1 85623 295 1

Praise About the Book

"Why have an orchard – even in your own back garden? The why, the how and the stories to get you going from someone who's answered the question in his own small back garden and then in a field he bought at the edge of his village. Makes orcharding seem possible even for us gardening fools. I love this book."

Sheila Dillon
The Food Programme, BBC Radio 4

"I feared this book would boost my yearning for an orchard. It did, delectably, and gave so much more. Their rich history – from Romans and local traders to expansion and intensification – is fascinating and the reams of invaluable, expert advice, images and clear love for these vital foods, trees and habitats is irresistible."

Vicki Hird
Head of the Sustainable Farming
Campaign for Sustain

"On a scale likely to be relevant to most of us; from the garden to 1 acre field orchard, this reads as an enjoyable and comprehensive 'step-by-step' guide, full of humility and infused with a passion for nature, whilst lacking in unnecessary jargon and any sense of ego. Wade makes the case for a UK fruit-growing revolution – never has this been more crucial as the whole of society must shift towards carbon-negative systems. Permaculture principles provide the guiding framework and what makes this unique is its spirit of experimentation and acceptance based on the author's own protracted observation; despite the best of our design efforts, nature will ultimately decide what will and will not work."

Lewis McNeill
The Orchard Project

"Covering all stages of creating and enjoying your orchard, from bare ground, through propagation and your first harvest, to managing old veteran trees and, most importantly, taking the time to sit and reflect, this clearly written, accessible guide is an enjoyable and engaging read, each paragraph interwoven with pearls of wisdom gleaned from the author's broad experience."

Steve Oram
People's Trust for Endangered Species

"Wade Muggleton makes growing fruit fun. Our trees go far when outrageous diversity and fungal precepts are an integral part of orchard design. Tracking down exceptional varieties will draw you in to apple lore and plum revelry. Clear instruction in horticultural skills like grafting further broadens horizons. The overarching refrain of *The Orchard Book* is that you can do this!"

Michael Phillips
Author *The Apple Grower* and *The Holistic Orchard*

"This is all you need to have an orchard whether you have a small space for a few trees or are planning to grow commercially. All, and more besides. Inspiring but also hugely practical. It gave me a feel of being able to make an orchard, a feeling of wanting to make an orchard and a feeling of the continuity of orchards throughout history. The illustrations are perfect, from small gardens to paddocks and everything else on the way. You'll not only plant an orchard, you'll enjoy the journey and find a community in this book. You can even make cider, vinegars and preserves. This, like an orchard, is a book that keeps on giving: giving confidence, ability and community. I loved it and will refer to it throughout my smallholding life."

Liz Wright
Editor of *The Smallholder* and
Smallholder Member of the Guild of Food Writers

"This is such a wonderfully fascinating and beautifully illustrated book on orchards and fruit tree growing, it is hard to know which part to highlight! Wade Muggleton's comprehensive research and experiential knowledge allows him to cover all aspects of growing apples and other fruit, from a practical viewpoint, as well as the wider issues of orchard history, community involvement, personal wellbeing and species biodiversity. In a world where more than ever we need sustainable ways of sharing the land with other life forms alongside getting an edible harvest, I believe this book will encourage, inspire and guide those who read it to realise this goal in whatever size of plot or land is available to them."

Jane Gleeson
Former Head Gardener and Programme Lead
for Horticulture at Schumacher College in Devon

"*The Orchard Book* is a wonderful introduction to orchards and their significant importance in the culture of parts of the UK. Seeing a revival in orchards and how they have become not just places that feed us, but vital in the fight for both climate and biodiversity is a joy, but adding in the importance of these spaces for community cohesion, just shows that all aspects of community food production can really support communities to find local solutions to global issues."

Sara Venn
Food and Growing Activist, Writer, Disruptor

"*The Orchard Book* is such a fruitful read from the history of apples to why and how to create an orchard. It's written in detail, but without complication, and takes a lovely holistic approach with permaculture at the heart of everything. I also enjoyed the quotes throughout which are a warm reminder of the important of fruit and nature."

Ellen Mary
Horticultural broadcaster, podcaster
and author of *The Joy of Gardening*

About the Author

Wade Muggleton has been interested in permaculture for over 30 years, having travelled in Australia and New Zealand in the 1990s where the concept originated. He has subsequently been a college lecturer and a countryside officer in local government. He lives in Shropshire with his partner and two children, where their plot, Station Road Permaculture Garden, has been a demonstration site for permaculture as well as open under the National Open Gardens Scheme. In 2013, he acquired a field which allowed his orchard ambitions to expand. He now has a collection of over 130 varieties of fruit trees and was featured on *BBC Gardeners' World* in 2018. He has written for *Permaculture* magazine for many years and regularly teaches, gives talks and runs training sessions on a wide range of orchard subjects.

COVER ILLUSTRATION by Sarah Robinson

Sarah trained in Illustration and graphic design in Bath in the early 1990s and later took a course in linocut which sparked an enduring love of the craft. She also enjoys working in coloured pencil, acrylic paint and line.

She lives with her family in the picturesque market town of Shaftesbury in North Dorset where she can often be found walking the hills and lanes around her home. Her much loved, but not very expertly kept garden, and also the wider natural world provide the inspiration for her work which is also informed by her love of folk tales, history and 20th century British artists.

www.sarahrobinsondesigns.co.uk

Contents

Introduction ix

Chapter 1 – What is an Orchard? 1
The history of orchards 3

Chapter 2 – Why have an Orchard? 8

Chapter 3 – The Fascinating World of Apples 12
The history of the apple 12
The scope for variety 14
Ten great apple stories 21

Chapter 4 – What do you want from your Orchard? 27
Six key questions 27

Chapter 5 – Exploiting your Space 32
The garden orchard 32
The field orchard 38
The inter orchard – the opportunity for so much more 42

Chapter 6 – Finding your Perfect Combination 47
How to select the right varieties for your needs and space 47
Pollination 49
Diversity 52

Chapter 7 – Going Beyond Apples 54

Pears 54

The plum 60

Other more unusual top fruit 63

Chapter 8 – Sourcing your Trees 68

Where to source 68

Seedlings 71

Grafting 72

Chapter 9 – Making your Planting Plan 78

Access 79

Orientation 79

Trees at altitude 80

Protection 81

Forms to consider 85

Chapter 10 – Giving your Trees a Good Start 91

Tools of the trade 91

Ground preparation 92

Planting 93

Weed control 94

Chapter 11 – Dealing with What Lies Beneath 97

What are you going to do with the grassland beneath the orchard? 97

Chapter 12 – Enjoy the Harvest 101

How to pick 101

How to store 102

Processing 103

Selling 107

Chapter 13 – Managing for the Long Term 110

Applied observation 110

Pruning 112
Mulching / Mucking 119

Chapter 14 – Working with and for Wildlife 120
Build it and they will come 120
Wildflower margins 123
Bees 124
Larger wildlife 125

Chapter 15 – Joining the Community 128
Apple day 130

Chapter 16 – Deeper into the World of Apples 132
Orchard detectives 132
Into the future 137

Epilogue 141

Glossary of Terms 143

Bibliography 144

Useful Sources of Information 145

Acknowledgements 146

Index 147

Introduction

Orchard – the very word has a somewhat exotic, almost mythical element to it, conjuring up images of beauty, richness and productivity. Be it the apple orchards of western England, the citrus groves of the Mediterranean, or the mythical lost gardens of Persia, there is something captivating about the notion of cultivating fruit trees that give their bounty in such rich edible forms. To walk beneath your own trees and pluck a fruit straight from the branch is an aspiration many of us hold.

Our earliest evidence of serious fruit growers here in the UK is the Romans, who had Pomona as their goddess of fruitful abundance. Derived from the Latin *pomum* meaning 'fruit', as such, she was the goddess of fruit trees, gardens and orchards – a fitting deity perhaps for permaculture itself.

Our relationship with fruit trees is long and filled with many stories. It is said that when those Pomona-worshipping Romans first came to the shores of England, what really impressed them was the richness of the seas around our coast, but after residing here for 300 years, what impressed them most was our climate's ability to grow fruit.

Now 1800 years later and with the Romans long gone, for all the rain and grey days, Britain is still a wonderful place for the cultivation of orchards and the beauty is that whatever our situation and at whatever scale, we can all have an orchard and be part of this continuing tale. From a couple of trees in a backyard, to allotments, gardens both small and large, through to whole fields, it is really just a case of good design and a few key skills, and that bounty can be yours.

Where I hope this book differs from the countless 'How to Grow a Fruit Tree' tomes is that I am looking both into and beyond the orchard, integrating a range of other elements that lead to something so much more than the standard dictionary definition of an orchard, i.e. 'an enclosed piece of land with fruit trees on it'. Trees are of course the key structural element in any orchard but need not be the only product; with good design we can integrate a range of yields and cropping opportunities as well as greatly increase biodiversity.

I don't fully recall the exact point at which my own personal journey into permaculture began. It was sometime in the late 1980s, possibly in the pages of *Resurgence magazine*, but it was seeing the documentary *In Grave Danger of Falling Food** when it was first broadcast in the UK in 1990 that was a light bulb moment. I was studying Rural Resource Management at the time at an esteemed but very traditional agricultural college and listening to Bill Mollison made me question almost everything we were being taught. It did truly change my view and in some ways the direction of my life. After college I travelled in New Zealand and Australia, visiting permaculture projects largely through the Willing Workers on Organic Farms (WWOOF) scheme. On returning to the UK, I taught in a regional college of agriculture and horticulture and, whilst permaculture was never on the syllabus, it continued to shape my thinking.

It was not until 2002 when I purchased my first house, and more excitingly garden, that I was for the first time able to implement my ideas in a literal and practical sense. That in turn led to 'Station Road Permaculture Garden' which, 18 years later, is still developing as an example of how the 80ft by 40ft (24m × 12m) plot behind an ex-local authority semi has turned into a micro farm.

Having built the garden it went on to become a demonstration site through the Permaculture Association's LAND project, being visited by individuals and groups from near and far. Thinking the confines of the garden's fences would be the limit of our permaculture experimentations, it was in 2013 that a field in the village came up for sale and we found ourselves the joint owners of 11 acres (4.4 hectares), and so our permaculture journey took on a whole new and greatly expanded dimension.

One acre of that field has been converted from overgrazed sheep paddock to my permaculture orchard, home to over 130 varieties of fruit trees and a range of other experiments.

This is the story of that journey told on two fronts: initially that of the garden and how to grow a domestic permaculture orchard on a very ordinary plot, and then taking it a significant step further, using the knowledge and lessons learned in the 'back garden' system to apply it on a far bigger field scale. In these pages I have tried to capture my experiences and ideas as to what an orchard can truly be.

Planting, nurturing and tending your own fruit trees on any scale is a wonderful and life-affirming experience; to walk out the door and pick your own, fresh fruit is a truly wonderful thing. Whether you are an existing orchard owner looking for diversity or a 'newbie', I hope you find something in the following pages that may also work for you.

---

* The documentary *In Grave Danger of Falling Food* was first shown in 1990 as one of a Channel 4 series 'Visionaries'. Although rather dated now in production terms, the content is as valid as ever and it can be viewed online for free on various platforms.

It is remarkable how
closely the history of the
apple tree is connected
with that of man.

Henry David Thoreau,
Wild Apples, 1862

CHAPTER 1

What is an Orchard?

So what actually is an orchard? We probably all have an image that springs to mind but for each of us it will be a slightly different picture: fruit trees of some sort, perhaps in neat rows, perhaps something more rambling and dishevelled, with bountiful blossom in spring and abundant fruit in the autumn?

From the diversity of an abandoned traditional orchard bristling with wildlife, through to the ultra modern commercial production in heavily sprayed intensive bush orchards, there is a great range of what we might call an orchard.

The modern commercial orchards can seem almost like the factory farming of fruit trees: intensively planted, heavily cropped for a few short years (using sprays and fertilisers) and just as quickly grubbed out – often when a newer variety has found commercial favour in the market. All of which is a very long way from the traditional and romanticised images of rosy-cheeked rural folk picking bright red apples from huge trees in the autumn sunshine.

If these are the two ends of the spectrum, then there are also many different sorts in between. We certainly can't claim all orchards are remotely the same, nor that they are all wonderful places. The silence in some modern orchards tells its own tale of them not being particularly desirable places for people or wildlife.

At this point we should perhaps define what we mean when using the term 'traditional orchard'. A traditional orchard is one where large standard trees are grown; they would typically have a 100-year plus lifespan and many such orchards would have had sheep or other livestock grazing beneath the trees at certain times of the year. The trees were deliberately grown on high vigour, or 'standard' rootstocks, which allowed for these tall trees, with a crown 6ft (1.8m) or more from the ground that allowed people and livestock to move easily through the orchard. The wide spacing allows in enough light to grow a good crop of grass for the grazing animals, making it a dual cropping system of the trees and sheep.

Apple trees with sheep grazing beneath, perhaps many people's notion of an orchard

In recent years, the remaining traditional orchards have become recognised as a priority habitat for the richness of our native flora and fauna with some suggestions that they are among the most diverse of all UK habitats, but there is a dilemma, some might even say a contradiction here, in that their abundant biodiversity is now the product of their abandonment.

In their heyday many of the traditional orchards were intensively managed and the fruit production was seen as a serious business to the extent that the trees were pushed hard to produce their fruit. The use of sprays, the removal of deadwood and unproductive trees, and pest control which often included killing birds, were all pretty normal practices. It is arguable that the decline of such management systems is what has been the making of them as the wildlife habitats we see today.

The dilemma this presents us with is that we cannot easily recreate traditional orchards and, even if we could, there is simply no money in doing so. Orchards historically served a purpose – fruit production – and any wildlife in the orchard was merely a by-product of the system. The demise of traditional orchards is the result of several factors: changes in wider agriculture; introduction of bush orchards where fruit can be far more quickly, easily and safely picked from the ground whilst big trees, picking using long ladders, and an abundant rural workforce, are all facets of a bygone age.

It is now, many years later, we recognise that the demise of traditional orchards has allowed them to become the rich and biodiverse habitats that we should cherish. As permaculturists we can, and surely must, develop a half-way house of creating productive orchards that also have as high a degree of biodiversity value as possible. (This will be explored in Chapter 14.)

Learning from the past, we must now look to the future and our moral obligation should be to produce homegrown fruit instead of shipping it halfway around the world with an enormous carbon footprint. Our local British orchards should once again become productive in a similar way to times past, supplying healthy British and homegrown produce. We should not consider them as mere nature reserves and havens of wildlife for that was not their original purpose.

Orchards, their history, their rise and subsequent decline, are interwoven with our culture both national and local, something we have rather lost sight of. If we cross the channel to Normandy and Brittany we would observe how, despite the devastation of two world wars, and more, the French have maintained their orcharding culture in a way we should be most envious of, for we in the past had one of the richest orchard cultures anywhere on earth, yet sadly one we allowed to slip away.

THE HISTORY OF ORCHARDS

Apples have been grown in Britain for thousands of years (see Chapter 3). Archaeological records as far back as the Neolithic and early Bronze Age show that apples formed part of the diet of early Britons. There is no evidence, however, to suggest the specific cultivation of fruit trees and the apples that were eaten might well have been gathered from wild growing apple trees. It is not until the time of the Romans that we start to get a notion of a specific horticultural approach to the growing and tending of fruit trees, and the birth of an orchard culture in Britain.

Whilst records of Roman orcharding in Britain are rare, there is documented evidence that Roman garden culture in Italy was a serious feature of their daily life. It therefore seems highly plausible that the Romans would have brought this growing culture to these shores and practised it in the centuries they resided here as they brought many other examples of their home comforts with them.

Orchards as we know them today have no doubt existed for many centuries, possibly since Romans times, but as a subject they are not well documented in records and literature, though they are sometimes seen in historic paintings and drawings. Those that are recorded are the orchards of the grand houses and estates that would not have been representative of the majority of the 'ordinary' orchards found on farms and in rural villages.

This lack of evidence is a reflection of the way in which so much rural history and especially that of the poorer folk was simply not written down, possibly due to low levels of literacy amongst rural people and also probably due to the view that it was simply not deemed important or worthy of recording.

Historically, the everyday orchards were likely small affairs geared towards supplying fruit for local use. Fruit is a heavy, bulky product and some fruits like plums do not travel well or keep for long once picked from the tree; for most of our history, distribution was just not feasible on a large scale or across great distances. It was the coming of the railways in the mid to late 1800s that changed the situation dramatically. With the increased transportation opportunities that the trains offered, entrepreneurial landowners saw a new business emerging; they planted large orchards near the new railway lines and thus accessed markets never before available to them. Fruit could be picked in the morning, put on the lunchtime trains and be in the markets of the big cities the following morning, something never possible until then.

Left: With a crown of 6ft (1.8m), the tall (standard) trees allowed people and animals to easily move beneath them

Below: Modern bush orchards are a completely different entity, short trees and short lived

In many places orchards and fruit trees were key features in the landscape; this ancient cherry is splendid with autumn colours

Orchard expansion was rapid and for a generation or two the industry flourished. Older Ordnance Survey maps are one way to observe how orchards came and went in the landscape. They are very different to, for example, ancient woodland, which may exist for centuries on a single site, whereas orchards can be planted, exist for a few decades and then be gone as quickly as they came.

Studying maps from the 1890s through to the 1930s, and comparing them to the same area's maps today, shows the scale of orchard loss in many parts of the country. Some districts were once almost wall-to-wall orchards; fruit growing was the main industry in many villages and towns, resulting in a landscape characterised by fruit trees. The blossom spectacle in such places must have been an extraordinary sight to see in their heyday and one we cannot begin to imagine in the 21st century. The Worcestershire town of Pershore, for example, had a huge industry built on the acres and acres of plum orchards that existed in the Vale of Evesham.

Across vast swathes of the country many, perhaps even most, farms had an orchard that was an integral part of the traditional mixed family farm of crops and animals. The orchard was a good example of dual cropping with fruit gathered from the trees whilst the animals grazed beneath – eating the grass and fertilising the trees at the same time.

An orchard can equally be a few trees in a domestic back garden

A large number of farms also made cider and perry for their own and their labourers' and workers' consumption. Thus the orchard was literally at the heart of the farm, producing crops of lamb, wool, beef, fruit and alcoholic beverages.

The various seasonal elements of the orchard year were an integral part of many rural people's lives. Picking provided seasonal employment; the income from the harvest season would often be a considerable proportion of the farm's total annual income, whilst pruning, cider making and even bird scaring were all important jobs on the agricultural calendar.

After the Second World War, as farming changed and the small mixed family farms gave way to larger specialist monoculture production of cereals, dairy or sheep ranching, the orchards declined. Indeed, some estimates suggest 90% of traditional orchards have been lost since the 1950s and with them has gone much of the landscape and wildlife associated with them. Some would say the seeds of decline had already germinated much earlier, even as far back as the late 1800s and into the early 1900s when large scale fruit importation from the New World was well underway; the notion of homegrown was perhaps already slipping away as much as 150 years ago, so setting in motion a slump in British orchards from which we have never recovered.

The two world wars totally changed the nature and make-up of the rural workforce, with vast reductions in the numbers of those working on the land. The decline of the railways combined with the parallel growth in road haulage meant it was no longer necessary for orchards to be located in close proximity to the railway network. Then, in 1973, Britain joined the European Union; the doors were opened unleashing a flood of cheap imported fruit onto the market. It was for many the death knell of an already mortally wounded industry. The grubbing out of orchards was subsidised by government policy and so huge swathes of orchard disappeared from the rural map.

However, what is gone is gone. Rather than lament past losses we should now focus on what we can create. The fact that fruit is a bulky and heavy product, which is also fairly fragile, completely justifies the permaculture principle that we should grow food where it is needed. The transportation, energy consumption and carbon footprint costs of moving fruit around the world are enormous, as opposed to growing it in towns, gardens, allotments and smallholdings where it will be more a case of food metres rather than food miles. Through good design we could incorporate fruit trees into all manner of our human spaces, be it parks, streets, gardens, the grounds of public buildings, factories and offices. The potential is vast; we just need the vision and the will to make it happen.

An orchard can be any size and any scale: two, three, four or more trees in a garden or allotment or 100+ trees at field scale. I have done both and in the chapters that follow we will trace those parallel journeys looking at the similarities and the differences of scale.

They (trees) consume
little from the earth and
indeed give back much
more than they take.

Richard St Barbe Baker,
1970

CHAPTER 2

Why have an Orchard?

As we all have different images of what an orchard is, we will also have a range of reasons for wanting to create or own one. Aside from our personal preferences, there are good planetary reasons for sustainable local orchards; by attempting to grow fruit where it is needed we will help to reduce the vast network of transport, packaging, pollution and waste that all too often goes hand-in-hand with not growing locally. You may simply wish to pick some fruit for you and your family or you may have grander aspirations to produce for others and distribute more widely to your locality.

Planting fruit trees can store carbon; provide blossom which in turn provides important nectar for pollinators; the trees will attract birds, encouraging them to build nests, and provide shelter and shade to people and animals. Environmentally trees help to stabilise soil by increasing its water holding capacity and aid natural flood management through increased infiltration; they also produce oxygen and of course grow that wonderful homegrown fruit.

Aside from the serious planetary and environmental reason to be 'Growing Our Own', another joy of having your own orchard is being able to cultivate and harvest a range of fruit that you could simply never buy anywhere as fresh produce. Of my collection of 130 varieties, all bar five or six would never be seen for sale in any shop; to be able to pick fresh fruit from the tree that you simply can't buy anywhere is a unique and wonderful thing. You can have a historic variety with a great local tale (see Chapter 3); grow a piece of living local history, something distinct to your village, town or district; grow the weird and the wonderful, the colourful, the quirky and the downright delicious. There are hundreds and hundreds of varieties to choose from and you can create an orchard large or small like no other, one that is totally unique to you.

Horticulture aside, an orchard can be the very essence of mindfulness, a place to lose yourself away from the stresses and strains of life. As I write

Left: Beautiful blossom
for humans, for insects and
ultimately for fruit

Right: Having your own orchard
means you can grow the unusual
and the rare, like these stunning
Worcester Black Pears

Above: Not all fruit need
be totally productive;
these 'Butterball' crab
apples add colour to any
garden or orchard

Right: What could be
better than homegrown
fruit right outside the
back door?

this in the midst of the COVID-19 pandemic of 2020/21, it is perhaps never more poignant. To simply be among your fruit trees is good for body and soul. In early spring when the blossom is at its most glorious, taking a few minutes to stand and observe reveals how such a simple act as planting and nurturing a few fruit trees is to be part of something far bigger. The number and variety of insects and birds that will come to your tree(s) is remarkable, key parts of an ecological system of food chains, feeding, breeding, life and death that you have aided by the simple act of tending a few trees.

Orchards are about so many things, fruit production yes, but history, landscape, wildlife and culture. I was once walking around a Worcestershire farm which still maintains its glorious ancient orchards, when the owner Peter and I stopped by one of the true veteran apple trees with its top long ago blown out and hollowed out trunk. Peering into a large cavity in the trunk, we came eyeball to eyeball with a barn owl. I am not sure who was the more surprised, the barn owl or us!

In 25 years of involvement with orchards, I have come across so much more than merely fruit and trees. I have found historic remains that hint at the orchards' past lives, amazing plants and animals indicating that traditional orchards are one of our most diverse habitats, unexpected encounters like with the barn owl, fascinating owners with priceless tales of orchard life, communities rallying around to save their orchard culture, fellow fruit enthusiasts, eccentric cider makers and so much more. From the large to the very small, from city allotments to rural Somerset or Herefordshire, from a few trees in a garden to several hundred majestic veteran trees in a centuries old orchard, there really is so much to orchards.

Why wouldn't everyone want one?!

Occasionally old orchards still retain historic artefacts from their heydays

WHY GROW YOUR OWN FRUIT?

In the age of supermarkets and global distribution networks you may well wonder why you would want to bother growing your own. After all, you can buy fruit reasonably priced on every day of the year. Well, there are a host of reasons to grow your own.

Freshness – when you walk out the door and pick homegrown apples, pears or plums straight from the tree, you will not get fresher fruit anywhere else. Some supermarket fruit is weeks or even months old by the time it makes it into your shopping basket.

Carbon reduction – by stepping outside and picking an apple from your own tree as opposed to buying one from New Zealand, there will obviously be a significant difference in the carbon footprint of that single apple. A case of food metres rather than food miles.

Waste and packaging reduction – again picking your own from just outside the window means the whole global distribution system, with its packaging, transport and waste, is avoided all together.

You know what has happened to it between picking and eating it – whether you are organic or do some spraying, either way when you eat your own, you know exactly what has or has not been involved in its production. However, with bought produce, you have little, if any, idea what it may have been treated with or what other factors have been involved in its production and route to you.

Biodiversity – by having your own trees that grow, blossom and then fruit you are contributing to the planting of trees in your locality and to providing habitat for local wildlife, large and small.

Grow something different – the range of varieties in the shops is extremely limited, but by growing your own fruit you can harvest and enjoy varieties which will never ever be found in the shops. Don't grow a tree of a common widely available type; grow something amazing and different, and preferably one which is local to your area

And lastly – just enjoy it. There is a wonderful sense of satisfaction and enjoyment to be had from planting, tending and harvesting from your fruit trees.

CHAPTER 3

The Fascinating World of Apples

Apples may be interesting but where exactly did they originate?

THE HISTORY OF THE APPLE

The last Ice Age made the British Isles something of a blank canvas in terms of vegetation. When we talk about where plants come from and whether they are native or not, the reference point therefore is whether they were growing in Britain at the point when it became an island and separated from the rest of mainland Europe, some seven thousand years ago.

This is a somewhat recent timeframe in the history and evolution of plants, but a defining moment in the history of Britain. The apple was undoubtedly here at the point the English Channel came into being, flooding the land bridge to the rest of Europe and making Britain an island.

The apple that was here then and is still here now is *Malus sylvestris*, the wild apple, and until comparatively recent times it was always and logically assumed that the larger eating and cooking apples with which we are familiar had been selectively bred or improved from their smaller, bitter, wild ancestors. This, after all, is how most agricultural plants and domestic animals came into being, through selection and breeding.

It was only around 30 or so years ago through developments in DNA and genetic fingerprinting, together with access to fruit from the formerly closed eastern countries, that it was discovered that the large cultivated apples that we eat and grow today were not closely related to *Malus sylvestris* but were largely genetically derived from wild apples that grew further east.

They originated in the mountains of Kazakhstan and western China; there forests were discovered where apples grew which were remarkably similar to *Malus domestica*, the cultivated apples we grow commercially. The science showed that the apples in those forests were the genetic ancestors of the apples in our fruit bowls – so not descended from *Malus sylvestris* as had been previously thought. They must have made their way west down the early trade routes and with the migrations of people. Modern science therefore showed without doubt that the origins of cultivated apples lay in the east and not from the smaller wild apples we often see growing in the woods and hedgerows of the UK. We now know that hybridisation is ripe, so today many apples contain a mix of genes from cultivated fruits as well as a sprinkling of genes from wild apples. In fact the true wild apple, *Malus sylvestris*, may be under threat in its true genetic form for such is the level of cross-pollination that its offspring are most likely a mixture of cultivated and wild apple genes all blended together.

Pears, the genus *Pyrus*, are believed to have a very similar ancestry to their cousins *Malus* the apple, also with origins in the east. Whilst the wild apple and wild pear have sat here in the landscape doing their own thing for six or seven millennia, their cultivated forms are incomers and inseparably linked to human activity, to people, to travel, to agriculture, to tree cultivation and to orcharding. The actual arrival on these shores of domestic apples, pears and plums is lost in history, but the Romans undoubtedly introduced or reintroduced them and certainly grew them around their villages and villas, setting in motion a stream of importation and introductions that, as we shall see, continues today.

The climate of Britain during Roman times is believed to have been different, being milder and possibly wetter. As far north as Wroxeter in Shropshire, they were able to grow citrus alongside apples and pears in a way not possible today.

Science and DNA shows the cultivated apple (left) isn't closely related to the wild apple (right) after all

Monasteries are also known to have played a key role in the history of fruit cultivation in Britain; records show all the major monastic sites had orchards. We can merely speculate due to lack of recorded evidence but it is likely that any fruit cultivation, which continued after the fall of the Roman era and through the Dark Ages, is probably the result of monks preserving the skills of orcharding through those barren centuries until such times as an interest in horticulture was rekindled in mainstream society and carried through to today.

The fruit stock of Britain was no doubt continuously added to by invasion and exploration. The Normans would have brought their preferred varieties across the Channel and the Tudors were known to be keen fruit growers, with Henry VIII sending representatives over to France to acquire the best varieties to bring back, graft and grow in the grounds of his large estates. In doing so, he improved the orchard stock of the country by adding the best that was available at the time from across the channel, a trend that has continued ever since with new and improved varieties being constantly brought to the UK from all over the world.

THE SCOPE FOR VARIETY

Having ascertained that wild fruits have existed here for millennia and that their cultivated cousins are more recent introductions, we are then led to the question of where all the specific, named varieties of fruit come from and how there are so many of them.

Some varieties are chance seedlings stumbled across by mistake, growing wild in a verge, hedgerow or some forgotten corner, a seed that germinated, got away and produced a decent fruit that someone thought worthy of propagating by grafting. This is how until comparatively recently all varieties came about, just randomly. It was not until the head gardeners of the Victorian and Edwardian era that we see a change and attempts to specifically breed fruit became popular. To create a new variety is, like any livestock or plant breeding, based on taking two good parents and crossing them in the hope of producing a greater progeny. This is the notion that by taking the best from both parents one comes up with something even better. Whilst the chance seedling 'find' has unknown parentage, the bred apple or pear is a more scientific calculation, selecting parents for various characteristics and then crossing them in a controlled manner. Today, the quest to breed the next big commercial apple, pear or plum is a high stakes business. Horticulture and agriculture have long embraced the practice of deliberately breeding livestock of note in the hope of developing an even better breed that becomes a commercial success. Such has been the drive to change or 'improve' the animals, fruits and crops grown, that those we see today are very different from those which populated the same farms and gardens two or three centuries ago.

My daughter with a selection of varieties from our collection

This practice of fruit breeding is carried out by selecting two parent trees with worthy characteristics and then keeping them in a controlled location so when in blossom, the flowers can be deliberately cross-pollinated by hand, thus ensuring an exact cross. The controlled environment (a bio-secured greenhouse or laboratory) ensures that insects are unable to come in and carry out unwanted pollination with unknown pollen.

The flowers are then carefully labelled with the details of the cross and in due course the fruits collected will be of that known parentage. Once harvested, the seeds within those fruits would be sown, germinated and grown on for a number of years until they produced fruit of their own. This would then be evaluated for flavour, keeping quality, disease resistance etc.

After many years of waiting, observing and evaluating, it is a sad fact that most are discarded as nothing special, but occasionally that huge effort would have paid off and a new variety worthy of marketing was created. It can then be named, propagated and commercially introduced to the world.

A few notable apple examples of this sort of breeding include: 'Merton Worcester', which was bred in 1914 by crossing 'Cox's Orange Pippin' with 'Worcester Pearmain'; and 'Lord Lambourne', a UK favourite in the 1950s, which was a cross between 'James Grieve' and 'Worcester Pearmain', bred in 1907. A slightly more modern example is 'Falstaff', raised in 1965 by crossing 'James Grieve' with 'Golden Delicious' at the East Malling Research Station in Kent.

From just these three examples a pattern emerges. Certain varieties like 'Cox's Orange Pippin', 'Golden Delicious' and 'Worcester Pearmain' have been crossed over and over again with all sorts of other varieties to try and enhance the flavour of Cox, the early season of 'Worcester Pearmain', or to introduce the keeping quality of 'Golden Delicious' into the new variety. Thus it is obvious that the gene pool of bred apples is becoming more restricted when compared with random seedlings whose parentage is dictated by the behaviour of insects rather than the aspirations of men and women.

The extraordinary *Malus* 'Roberts Red' with its purple skin and flesh, an example of the diversity of apples

Two stories show the contrasting sides of the seedling versus deliberately crossed apple story. Perhaps the most famous cooking (culinary) apple in the northern hemisphere is 'Bramley's Seedling', or 'Brailsford Pippin' as it should perhaps be known. For it was one of those chance seedlings grown in a cottage garden in Nottinghamshire by a young girl named Mary Anne Brailsford. It was only some decades later, so the story goes, that the cottage belonged to a Mr Bramley, the village butcher. A nurseryman friend of his asked if he could take some cuttings of his apple tree to graft. Mr Bramley supposedly said, "Yes, as long as you call it after me." Thus it went on to become the most famous of cooking apples and Mary Anne missed out on her place in pomological history.

Contrast this chance seedling with the supermarket favourite 'Gala', which was bred in Greytown, New Zealand by Mr J H Kidd – the result of crossing his earlier prodigy 'Kidd's Orange Red' with the famous 'Golden Delicious'. He bred what went on to be one of the most commercially popular eating (dessert) apples in the world. In the process, no doubt, Mr Kidd equally bred an incalculable number of utterly nondescript or indifferent apples, a class out of which 'Gala' stood out to be the star pupil.

Around the world the aim of breeding the next worldwide commercial favourite that will go into every supermarket is probably as likely as winning the lottery three weeks in a row. But for those who chase that dream, the stakes are high; to breed a new variety of merit today is to possess a commercial commodity of considerable worth. Varieties today can be copyrighted and royalties charged for their commercial production, so potential fortunes are there to be had if you could come up with the next 'Gala', 'Braeburn' or 'Pink Lady'.

'Gladstone' from 1868, believed to be a chance seedling find by a Mr Jackson of Blakedown Worcestershire

As for the two thousand or so named varieties grown here in the UK, they are a mongrel collection of apples; some are homegrown whilst others have been brought in from across the world with origins in a vast array of countries from Japan to the United States, Russia to Europe and, as we have already seen, New Zealand. Many have found that the climate in Britain is much to their liking, growing and cropping as well in these conditions as they would have done at home. Unfortunately other varieties have turned out to be less well suited to our weather and soils. Apple varieties are just like people; what suits some does not suit others.

What is happening today is exactly the same as hundreds of years ago. We are continuing Henry VIII's quest to add the best from across the water to the roll call in British orchards. The only difference though is that we are trawling the world, whereas Henry VIII only went to France.

But what actually makes apples interesting? After all isn't one apple pretty much like any other? That is a question I have been asked on numerous occasions. To many people apples are either red or green, merely perceived as just a healthy snack sat in the fruit bowl.

There is far more to apples than those shiny specimens we see on the super-market shelves and it is lamentable that most of us will only ever see five or six different varieties whilst perusing the grocery section on our weekly shop. There are well over two thousand different varieties of apple growing in Britain alone and as many as ten thousand worldwide.

Why so many? Well apples belong to the Rosaceae family, so they are relatives of roses, and their genus is one of the most genetically diverse in the

'Discovery' (c) is believed to be a cross between 'Worcester Pearmain' (a) and 'Beauty of Bath' (b)

plant kingdom. But putting the science aside, the apple fact that most people are unaware of is: if you take a pip from your daily apple and plant it, it will grow into an apple tree and it will produce apples, but those apples will not be the same variety as the one they came out of, they will be different. That is because the trees grown from each pip are the product of cross-pollination – it is genetically fifty percent the apple it came from and fifty percent the variety which pollinated it – thanks to the insects. The pollinating variety could have been a cooker, an eater, a cider or a crab apple. Thus when you grow an apple tree from a pip, you quite literally never know what you are going to get. On many occasions you will get something small and crab apple-like, a reversion to the *Malus* ancestry. Occasionally you will get something similar to the apple it came from, but not exactly the same. Most times you will get something in between – a small to medium, fairly nondescript apple of no great merit. To many people this seems extraordinary but it is actually obvious; it is exactly the same as breeding animals or for that matter humans. I have two children who are unlike each other, nor the same as me or their mother. It is just the same with apples: they take genetically from both parents and so you get something unique. Even five pips from the same apple will grow into trees that produce five different apples; every pip is utterly unique.

Give it a go; grow an apple from a pip and see what you end up with (see Chapter 8). It will be something new every time. Such is the diversity of their gene pool, that if you were to grow a thousand more apple trees from pips, you will never get quite the same apple twice. Just as each human being on earth is unique, so are apples grown from pips.

This means that were nurseries and growers to propagate all their apple trees from seed, each and every one would all be different and none of us could buy a tree of our favoured variety. Fruit trees are therefore propagated vegetatively by grafting (or budding) which is the seemingly miraculous process of taking a cutting from one tree and splicing it onto another young tree, for this is the only way to replicate the variety, ensuring an exact genetic copy. Thus, of all these thousands of different varieties across the world, there only ever was the one tree of each – the original one, the first and only seedling of its type, and every subsequent tree of that variety is a graft of a graft of that original tree. The older historic apples will therefore have been replicated time and again down the centuries through grafting, and it would probably be correct to say many thousands of times in some cases.

When you purchase a young fruit tree from a nursery or garden centre, you are buying something completely unnatural and artificial. Your newly acquired young tree is in fact two different trees joined together as opposed to a tree grown from a pip, which is a natural occurrence. When purchasing your tree, you should ensure that you know the exact make-up of the tree, that is the name of the variety you are acquiring and also the name / type of the rootstock the variety was grafted onto. This is important as it will determine the size your tree will grow (see Chapter 6).

On a long autumnal car journey you may have caught a momentary glance of those apple trees that grow along the verges of our dual carriageways and motorways laden with apples of all shapes, colours and sizes. These may well have originated from the apple cores thrown from the windows of passing cars. If you were to stop and collect a specimen of the fruit (not advisable on motorways), you would not find them in any book, nor be able to put a name to them. For they would all be those unique seedlings, or random 'wildlings' as we often refer to them, each one different.

Not only is the genetic ascendancy of the apple interesting, but the realisation that every named variety was simply a seedling that someone deemed worthy of copying by propagation. This tells a story of what someone considered important in terms of character, flavour and keeping quality. Equally the origins of each of those named varieties have a tale to tell: the story of where they first came from; the village, region or county of their origin; the individuals who came across them, found them or in some cases bred them. Were they once popular and have now fallen from favour? Or were they never more widespread than just a few parishes in a specific part of the country? Apples are a living history and to plant and grow your own orchard is to write another chapter in that fascinating history of humans and orchards.

Returning to the Romans, there are references to named varieties in the works of Pliny who named 20 different apples in a list in his first century *Natural History*. Whether any of these made it to Britain we do not know for there is no written evidence either way to prove or disprove. Whether any varieties that the Romans brought with them still exist in Britain today is again unprovable but in my view it is extremely unlikely. It would have required a high level of grafting and orchard husbandry skills to be used continuously through the turmoil of the Dark Ages and right through to modern times. What seems far more likely is that the fruits the Romans may have brought here have long been lost to cultivation and what we have today is the result of more modern introductions from various parts of the world alongside a huge amount of native and cross-pollinated seedlings which were replicated and then became named varieties.

Moving quickly through two millennia, an example of the changes globalisation and mass production have heaped upon the fruit industry shows that in 2018, half the apples that were actually grown here in the UK were of the variety 'Gala' and half of the remainder were 'Braeburn'. This means that three quarters of the UK-grown apples were just two varieties. This, in a country with a rich history of over two thousand varieties, is a stark illustration of the loss of diversity and the reason most of us only ever taste so few varieties.

The second question that people often ask me is "How come you know so much about apples?" My reply is along the lines of "Well everyone knows a lot about something; Britain has long been a hotbed of passionate, knowledgeable experts and specialist folk... so for some it's football, cars, TV soaps or even breeding budgerigars... we all have our subjects and I guess for me it's just apples." Countless variation in sizes, colours and appearances, the vast range of flavours and uses, social history, genetics, climate, environment and seasonal unpredictability, and the sheer satisfaction which comes from growing your own. That is why apples are interesting... at least I think so.

TEN GREAT APPLE STORIES

'Catshead'

This ancient cooking apple was known by this name in the 1600s and could well go back much further than that. It is listed in John Evelyn's 1662 book, *Sylva*, so was known at that time. It has an elongated, angular shape and is of a type considered to be a primitive form of apple known as Codlins. It is a good old-fashioned cooking apple that can be baked or sauced. It gets its name from supposedly looking like a cat's head, perhaps not an obvious comparison, but maybe a few pints of rough cider improve the resemblance.

'Court Pendu Plat'

This late season dessert apple has been claimed to date back as far as Roman times but from where this story originated is unclear and it is of course utterly unprovable. However, like 'Catshead', it also has records back into the 1600s. Its fruits have a rich aromatic flavour and it does well in frost-prone areas as it is very late to break leaf and produce blossom compared to most other apples – by the time it flowers, late frosts in most areas should have passed.

'King of the Pippins'

This excellent mid-season dessert apple has a wealth of names and associations, known by various names in different places. In Shropshire it is known as 'Stanardine' as well as 'Shropshire Pippin', in Worcestershire as the 'Princess Pippin' where it is said that when the young Princess Victoria visited the town of Tenbury Wells in 1832 she was presented with a basket of 'Stanardines' and was so impressed that a message came back from the palace asking if they could subsequently be known as the Princess's Pippin. At the National Apple Congress of 1883 when 1545 different varieties were displayed in an attempt to boost the British Apple industry, 'King of the Pippins' was voted the favourite of them all, beating the now more famous 'Cox's Orange Pippin' into second place. A crisp, richly flavoured apple, it was once extensively grown for the market.

'May Queen'

Originally raised by a Mr Haywood of Worcester and first recorded in 1888, it was introduced to the market by Messrs Penwill and won an RHS Award of Garden Merit in 1892. In 1920, renowned fruit man, Edward Bunyard, stated it was a neglected fruit of great excellence and he could not understand why it was not more popular as it was the perfect tree for small gardens, being especially suited to being grown as cordons or bush trees and being very hardy. A medium sized red dessert apple, it will keep well into the winter months and is an amazingly regular and heavy cropping tree that is not at all vigorous, so does not require an excessive amount of pruning, hence being well suited to small plots.

'Pitmaston Pine Apple'

It is believed to have originated in Herefordshire around 1785, but it is associated with the Pitmaston Nurseries in Worcester, who introduced it to the market and from whom it derived its name. It is the sort of apple you never see in the shops, being small, yellow and spotty, but to catch it just right in the season is to experience a flavour that is simply sublime. Described by Bunyard in 1920 as "a remarkable mix of honey and musk", it is a good garden tree and a great apple for children; young children often only eat half an apple thus wasting the rest, so these small ones are ideally suited for young mouths. As a variety it has a tendency to be very biennial, i.e. producing a huge crop one year and then effectively having a year off before throwing another huge crop the following year.

'Granny Smith'

This much maligned supermarket apple really did originate from a Granny named Smith. A chance seedling that grew in Ryde, New South Wales, Australia on the property of a Mrs T Smith, it was recorded as fruiting in 1868. Although not appearing in Britain until 1935, it went on to be a supermarket staple and is still widely grown in South Africa, New Zealand, Australia as well as parts of Europe. Its greatest characteristic is its keeping quality, for it will store for longer than virtually any other apple kept unrefrigerated, lasting until June in a good year.

'Scotch Bridget'

This late season cooking apple originated in Scotland in 1851 according to the books, where it was also known as 'White Calville'. However, in the Worcestershire town of Tenbury Wells, there persists a rather different story where the natives believed it was a local apple that had originated as a seedling in the garden of the Swan Hotel in the town and had been given its name after a Scottish barmaid named Bridget who had worked at the hotel. It is hard to date this story exactly, but it seems to have been around the 1920s and 30s and the variety was widely grown in Worcestershire's Teme Valley in the early 1900s, so it would have been a very familiar apple in the area. Pub talk maybe, but it is interesting how these stories arise and perpetuate. Part of its appeal at the time was its amazing keeping quality; perfect specimens will keep until Easter or even through to May, which would have made it a much prized cooking apple that could be used all through winter in the days before supermarkets and refrigerated storage.

'Discovery'

As an apple to pluck from the tree and bite straight into, there can be few, if any finer, early season eating apples. It is not an ancient variety, being raised

in 1949 by a Mr Dummer of Langham, Essex, and then introduced to the market by Matthew's Fruit Trees of Thurston, Suffolk. It was originally named 'Thurston' and renamed 'Discovery' in 1962; it went on to become a popular early variety for the market, as well as the home grower. It was bred by crossing 'Worcester Pearmain' with 'Beauty of Bath', so a case of taking two well-known early apples in the hope of getting an even better progeny and in this case succeeding wonderfully, for it is a far finer fruit than either of its parents. The trees are fairly hardy and the blossom shows good resilience to frost, making it a good choice for growing in cooler areas. It is an apple of good flavour with its creamy white flesh sometimes tinged with a pink bleed below the skin.

'Winter Banana'

Worthy of a place in any orchard for its name alone, it is also a stunning looking apple. The name sadly does not have any association with the tropical fruit; it is simply the vivid yellow colour that they share. Originating in Cass County, Indiana in 1876, in its native US it is considered a superb pollinator and was sometimes grown in with other more commercial varieties for this reason alone. Picking time is October and they will keep well up to Christmas and into the new year. When caught in the early autumn sunshine they are one of the most spectacular looking apples you will ever see hanging on a fruit tree.

'Doddin'

This little apple is thought to be unique to the Worcestershire town of Redditch. The apples grow on small bush-like trees and can be ripe as early as July. It is claimed in the town that during wartime, when sugar was short, children were given bags of the little apples as substitute sweets. Some people eat them whole, core and all. It even has its own society, the Doddin Preservation Society, formed in its hometown to make sure it did not become another lost variety.

A glamping pod in an orchard as part of a tourism business

Juicing or another fruit-based product may be behind an orchard venture

A bountiful crop of homegrown fruit might be another reason

"If you want to create
an apple pie from scratch
you must first create
the universe."

Carl Sagan

CHAPTER 4

What do you want from your Orchard?

Creating a new orchard or purchasing an existing one is an immensely exciting project but can also seem slightly daunting, so plotting out the process and asking yourself some key questions are necessary, long before a tree is purchased or a spade is put in the ground. Like all good permaculture, observation and planning is key to success. Knowledge of your plot in terms of its soil, aspect, microclimate, exposure to wind and rainfall will all help shape your plan.

Then ask yourself the following.

SIX KEY QUESTIONS

1. Why do you want an orchard?

We looked in detail at why to plant an orchard. I will reiterate: an orchard is a wonderful thing in itself and that alone might be a good enough reason to plant or own one, but fresh home produce, a small business idea, a haven for local wildlife, something nice to look at out of the window, can all be reasons that individuals choose in creating or purchasing an orchard. Of course with good planning and design several of those objectives can be met in a single orchard.

2. What do you want to get from your orchard?

Perhaps similar to the above question but a place to relax, a bountiful crop of homegrown fruit, a biodiverse habitat, beautiful blossom, a site to entertain, socialise, camp or glamp. Again, multiples of these aspirations can be met through good design.

What might you grow – apples, pears, stone fruit, nuts, exotics?

There are no set-in-stone, must-grows for establishing an orchard; the best advice is grow what you like and are likely to use. Apples are the most commonly grown orchard fruit here in the UK, but if you like pears, plums, apricots, greengages, quinces or medlars, then grow them. Consider what amounts of each you are likely to need or use and in the event of surplus, do you have an outlet or use for it?

3. What scale are you thinking: 6, 20, 60 trees?

This will be dictated by the space available and the level of your ambition, your budget and vision for the orchard. The planning system recognises six fruit trees as constituting an orchard (I have over 100) but really an orchard can be any size from one or two trees in a small yard to hundreds on a field scale.

4. What size of tree do you ultimately want?

Big old-fashioned trees where you can one day sit in the shade but need a ladder to pick, or more modest sized trees that can be picked and pruned without ever needing to take your feet off the ground? Again your site, the available space and your vision may also influence your choice.

5. What is your budget?

Whether we like it or not money is a key factor in life, and orchards are no exception. Even the most stringent attempt at a DIY orchard will still incur some expense, so having a budget at the outset is key and the size of that pot and how far you can stretch it will have an impact on the scale and make-up of your orcharding ambitions. The major costs – assuming you do your own labour – will be trees, tree guarding, tools for planting and maintenance and on-costs to do with mulch, mowers, livestock, etc.

6. How much time do you have to look after it?

There is a commonly seen car sticker, which informs us, 'A dog is for life, not just for Christmas'. I have often thought of producing one that states 'An orchard is not just for planting', for it requires a commitment of time, materials and even love.

In theory an orchard is, in the longer term, less work and time than a conventional vegetable or kitchen garden. Once planted you are freed from the annual round of sowing, pricking out, tending and watering that goes with a vegetable plot. Equally an orchard does require certain seasonal, if different, commitments. In the early days weed control is essential, and as it matures, pruning, picking, perhaps fertilising and feeding the trees can all be part of the annual maintenance regime (see Chapter 13). So if you have a very busy life and limited time, then building that consideration into your design will be better for you and the orchard. Perhaps have less trees in total but on bigger rootstocks or have a row of easily maintainable cordons? All solutions will be bespoke to your plot and your lifestyle, but try and design a system that fits into your life.

Answering all of the above questions fully and honestly will go a long way to shaping your design and ultimately giving you the orchard that works for you.

Alongside these questions take time to observe those physical aspects of your site; spend time looking at it in different seasons and in different weather

What size of tree do you ultimately want?
Something small and manageable
Or something bigger?

MYTHS ABOUT GROWING YOUR OWN FRUIT

There are a range of myths about how growing your own fruit is some strange and mysterious practice:

It is difficult

Trees want to grow; that is what they do, all you need to do is supply the right conditions and nature will do what nature does best.

Fruit trees take up a lot of space

As we shall see in Chapter 9 there are different forms of fruit tree for all and every space you can imagine.

You need lots of skills to grow your own

With a few basic techniques and the simplest of tools you can easily manage and care for a few fruit trees, as we shall see in the pruning section of Chapter 13.

It will be years before you ever get any fruit to harvest

On the smaller modern rootstocks you can be harvesting your own fruit from year two onwards.

So don't be put off by old myths and doomsayers; it really is within the reach of anyone to grow their own fresh fruit.

Beautiful blossom is one reason to have an orchard

conditions. Note where the prevailing wind comes from, where the frost is most severe and lies the longest. Does rainfall sit on the surface or freely drain away? Record all of this by making notes on these factors. Photographic or even video footage are all useful design tools in recording these aspects of your site. As in all permaculture design, writing it down, taking time to mull it over, making a sketch, tweaking it, mulling it over again is all part of the process. Technological tools like camera phones, tablets and spreadsheets are all useful to aid your design.

From all of this accumulated information and observation, come up with a design. This does not have to be anything technical; just a clear sketch on a sheet of paper can be sufficient or something recorded on computer or tablet for the more technically minded. From your observations, consider that if you have a frost pocket it is best to plant very late flowering varieties so they miss the last frosts. When it comes to wet ground, fruit trees tend not to like sitting in saturated ground for weeks on end, so if you have a wet area plant your trees on raised mounds so the roots can go down into the wet if they choose but also have access to drier, more aerated soil as well. If your site is exposed, a windbreak might be a good idea (see Chapter 9). Finally, really give some thought to how you move around the site and make sure, when established and far larger, that the trees aren't going to be in the way of you moving through the site as well as any other activities you have planned for your orchard. Having come up with a plan, you then need to choose your trees, selecting the varieties that appeal to you, as we shall explore in Chapter 6.

*"When is
the best time
to plant an orchard?
Ten years ago..."*

Anon

CHAPTER 5

Exploiting your Space

Whilst everything discussed so far pretty much applies universally, there are, as we have said, different scales of orchard, be it a backyard, garden, allotment or field. Yours will be bespoke and your orchard will be unique to you in so many ways.

THE GARDEN ORCHARD

When starting at a garden scale, what you plant can rather depend on what you have to begin with. You might be fitting a few trees into an already established garden or designing a garden from a blank canvas.

On a domestic scale, almost all gardens can incorporate one or more fruit trees; it really is a case of the right tree for the right space. Whilst the field orchard may have taller trees (what we might think of as those traditional fruit trees), in the garden and especially small gardens and yards, we can look at trained and dwarfing forms of fruit trees. I have often heard the lament, "I don't have room for a fruit tree" when in truth, through good design, a range of fruit trees can be included in almost any garden.

Our plot, Station Road Permaculture Garden, has 30 fruit trees in a rectangular plot of an ex-local authority semi-detached house. I describe it as a very ordinary plot, for red brick, three bed semis of this type (ours is 1950s) were built in their tens of thousands across the UK and often had a decent sized garden: ours has a 40 by 80ft (12 × 24m) rectangle at the back with a 40 by 40ft (12 × 12m) square in the front.

The idea of developing it as a demonstration site was very much based on the principle that if it is possible on a plot like ours, then it may equally be possible on the countless similar sized gardens across the land. The problem with gardens and horticulture is that we have a tradition of reverence for the grand and opulent, with the spectacular gardens of stately homes and National Trust type properties among the most popular of visitor attractions. Yet perhaps following a visit to one, when we return to the very ordinary plots surrounding our semis, terraces, bungalows and estate houses, the relevance can be lost and a feeling of deflation and a "Well what can I do with my little space?" then ensues. Hence the value of demonstration sites, especially small ones. Seeing what others have done and taking ideas, tips and inspiration is priceless. I always say "Nothing is new and all good design and practice is merely an amalgamation of observations and ideas taken from elsewhere", and there is absolutely nothing wrong with that.

When we purchased our home in 2002, it was just a lawn, with two established apple trees on one side and a line of conifers on the other side, so essentially a blank canvas and good from the point of view of not having to rip out a previous garden to build the one we wanted. There were, however, a range of materials left behind by the previous owners and 'up-cycling' and reuse were high on our agenda, so everything possible was reused and repurposed; old bricks and tiles, broken slabs, pieces of timber all became new features.

The 'May Queen' on rootstock M26 when first planted (left) and 16 years later (right), it is the centrepiece of the veg plot and definitely the right tree in the right place

Our garden in three zones: zone 1 is the lawn nearest the house, zone 2 is the central vegetable area, and zone 3 is the orchard / chicken are farthest away

Attempting to implement permaculture design principles, I viewed the garden as three zones phasing away from the back door. A patio, herb bed and small lawn area in zone one; raised bed vegetable garden in the middle section as zone two; and the bottom third of the garden as orchard, meadow and chicken area, zone three. The concept of zones, of placing elements of a design in the right place, is a big part of permaculture. So those elements that require the most attention occur nearest the house or homestead and those that require little if any attention are placed farthest away. This principle can be applied at any scale from a garden to a whole farm and if done well, makes for a more efficient use of space, time and effort.

The first four fruit trees to go in zone three were two on standard M25 rootstocks (rootstock explained on page 50) and two on the middle-sized MM106 rootstock. So on a micro scale this was more of a mini-field orchard with taller trees and wildflowers and hens beneath.

For zone two, the vegetable garden, I planted an M26 apple as the centre-piece (variety 'May Queen', see Chapter 3); being on a semi-dwarfing rootstock, it would not grow too large and shade the vegetables, and 17 years later it is exactly the right tree in the right place, forming a focal centre, being very productive and in no way impinging on any other aspects of the vegetable plot. One side of the vegetable plot is lined with three stepovers (see Chapter 9) and down one side of the path are cordon apple trees trained on wires (also Chapter 9).

Eighteen years on, our garden orchard breaks down into a range of trees of varying ages on different rootstocks and in different trained forms. We have six stepovers, eight cordons, two standards, eight semi-dwarfing trees, one espalier, five dwarf bush trees and a family tree with seven different varieties grafted onto it.

The design was not quite as per the permaculture manual, and has been added to and had extra trees shoe-horned in over the years. If you are starting from scratch, it is well worth taking the time and planning it, even if it delays planting by a few weeks, months or even a full year. Consider what you want fruit-wise, what might work best in the available space. Do you want something that visually looks like a traditional fruit tree or might a row of cordons or a fan or espalier against a wall work best for you? Do you nominate a specific fruit or orchard area or do you mix trees in amongst other garden features? Making a plan will help with what works best and where (see Chapter 9). Visit other people's plots, read books, look at photos, watch films and visit websites to get ideas and inspiration that can help you shape your plans.

Poultry

The options for animals in the garden orchard are more limited than for the field orchard, with poultry being the most obvious choice, providing eggs as well as pest control below the trees. We have kept hens, ducks and at times a combination of both over the years. Although we currently have six hens, I do prefer ducks; they have more character and personality than chickens and do not scratch and dig up plants in the way the hens do. But they do love a pond, however large or small, and then equally like a wet muddy mess, so if you like a clean and tidy garden, ducks may not be for you. Also ducks do not produce eggs on anything like the scale you will get from modern hybrid hens, so if eggs are a key factor, then hens are possibly the better option.

Some moths and fruit tree pests spend part of their lifecycle as grubs and eggs in the ground around the trees, so having hens scratching beneath makes a valuable contribution to organic pest control measures. Modern hybrid chicken breeds like Warrens are very tame and easy to handle, and do not go broody, so make for easy hen keeping. Rare breeds may be more attractive and contribute to the preservation of our heritage poultry, but are not as easy for the novice.

Ours live in an 8 × 12ft (2.4 × 3.6m) fox-proof run where they are safe while we are out. When we are home and in the garden they have full access to zone three, so getting more stimulation and environmental enrichment from the orchard/ wildflower area. As well as a good supply of eggs, they contribute manure and litter from their house to the compost system and also eat scraps, weeds and waste from the vegetable plot. So as part of a cyclic permaculture system, hens are a great component.

So whilst the garden isn't perhaps an orchard in a conventional sense, it is, in our case, 30 fruit trees in amongst soft fruit, vegetables, flowers, ponds, chickens, water harvesting, sheds and greenhouses. That is the beauty of permaculture: the integrated nature through the flowing and merging of the edges of one thing seeping into another. This does, perhaps, challenge our perceptions of seeing orchards as a monoculture, or at the very least, fruit trees and grass beneath. In a permaculture sense, our garden is an orchard with a lot more elements added in – something we will investigate further in The Inter Orchard later in this chapter.

Ducks in my view are the best garden poultry, if rather messy. But hens are a more reliable option when it comes to egg production.

Spring 2020 and the blossom shows how fruit trees are integrated throughout the garden

THE FIELD ORCHARD

In 2013 we obtained our piece of land at the other end of the village and so I began to establish the field orchard, a project on a much larger scale. Until 2013, the entire garden project was based on an assumption that it would be the extent of my permaculture project. The field totally changed the dynamic and with hindsight, had I known I would one day have the field, there would be less fruit trees in the garden, but Station Road Permaculture Garden was built as a micro farm, an attempt to eat as much as possible and to eat well from a small plot, something I hope it still demonstrates.

As such, my situation is an odd one; because I can go up the road and harvest fruit from 100 trees in the field orchard, the need for the garden orchard is questionable. However, in telling these two parallel stories I hope that readers may take what is suited to their situation and use it or be inspired accordingly. Obviously, most people are not lucky enough to have a field so the garden still represents what it was set up to demonstrate.

My field orchard is not a commercial enterprise, but a collection, a museum orchard of sorts, for among my 130 varieties are some incredibly rare and obscure apples that are found in only a handful of places in the world. Most people looking to set up an orchard are likely to use far less and far more familiar varieties, tried and tested staples that will deliver a good crop. Whilst some elements of the field and garden orchard are the same, there are also key differences. A large expanse of just grass is the ultimate blank canvas as far as orchard design goes. So the old saying that 'the only limit is your imagination' may be true.

Answering those six key questions in Chapter 4 will dictate much of the design of a field orchard, whether like mine it is a collection/hobby or you have a business in mind, in which case factors like ease of harvesting, access to extract fruit, proximity to track or road will all come into play.

Layout

My field orchard sits at 600ft in the Shropshire hills of central England and slopes gently towards the north-west, so not an ideal site for an orchard, but not too bad either. Siting an orchard is a balancing act between shelter and exposure. Whilst a nice secluded valley bottom might seem like a good location, it can have a number of problems, such as cold air or frost lying in the valley, and poor air flow; this can lead to disease problems as spores and damp air are not flushed from the site. So my site does suffer from exposure and the wind but air flow is good, cold air sheds away down the slope and so far disease problems have been minimal. In most situations, you will have to work with what you have, but if looking for an orchard site from scratch, then not too exposed, but not too tucked away in a hollow, is the best of both worlds.

The ultimate blank canvas, the site of my field orchard in 2012

Unlike the garden orchard, my field orchard was more designed. Since it was a blank canvas, I based my layout of what is a rectangular plot on using trees at 7m spacings and for the apples, I selected M111 as a rootstock. This is a large rootstock and one claimed to have very good anchorage, i.e. the trees don't easily blow over; the exposure of the site influenced my choice of rootstock, an example of environment influencing design decisions. The size of the trees is one of those key questions; selecting the right rootstock is essential (see Chapter 6). Small semi-dwarfing trees can go their whole lives without ever needing to reach for a ladder or a set of steps, something that may be key for you. If you plant a standard orchard while you are young enough, you may be able to sling a hammock beneath them in a few decades' time. Decisions will be personal but are key to the overall plan.

Planting a sizeable orchard is potentially an expensive exercise, so I opted to graft many of the trees myself. I did buy eight apple trees on M25 rootstock to create a bit of instant impact. I then purchased a bundle of 100 × M111 rootstock at £1 each. With a wanted list, I then went collecting scion (cuttings) from as many interesting trees as I could from friends' orchards and gardens, as well as receiving material from fellow orchard enthusiasts through the post. I ended up with 40 varieties of apple in that first season and grafted two of each, in order to ultimately get one of each. I then took the unusual decision, one I have not come across anywhere else, to plant one of each of the newly grafted trees straight out into their final positions in the field. This is opposed to the traditional practice

of growing them on in a nursery bed or pots and planting them out at a year or two old, my theory being that in a natural situation trees grow where they germinate and no young trees really like having their roots disturbed by being dug up and moved around.

So I put these tiny newly grafted trees, barely 6 inches tall, 7m (22ft) apart, in a big open windswept field. I then kept the second of each variety in pots, as back up, to replace any failures. Using plastic deer tubes over the new grafts created a mini greenhouse effect, sheltering and warming the young trees, as well as drawing them up the tubes so that by the end of the first season, most had poked out the top of the tubes, making a metre or more of growth. Of the 40 that went straight out into the field, only one died and was replaced with its potted twin. I ended up with 39 spare potted trees which were given to friends and members of the local permaculture group.

This idea of planting straight out into their final position certainly saves work and avoids root disturbance. I am told it is practised commercially in the Netherlands but I have not come across it anywhere myself. For most people in most situations, this will not be realistic and purchasing trees from a reputable fruit tree nursery will be the way to proceed (see Chapter 8). From a financial

1 The tiny newly grafted trees went straight out into a big windswept field

2 A small sprinkle of mycorrhizal root powder to help their start in life

3 A forestry deer tube acts like a mini greenhouse, sheltering the young graft

4 Small beginnings: the tiny graft at the bottom of its tube

5 By the end of the first season most of the grafts were out the top of their tubes

6 At 16 months we had good growth and form starting to take shape

7 By 2018 it was really beginning to feel like an orchard with well established trees

aspect, my trees cost me £1 each, and with a stake and deer tubes at around £3, meant I established 40 trees for less than £5 a unit – tree, stake and guard. Had I bought all my trees instead of grafting them, and the necessary stakes and guards, I would have been looking at nearer £20+ a unit.

It took only two years to pretty well fill my orchard on the 7m (22ft) grid. I had then theoretically run out of space and, having a strong interest in apples, I continued to come across other varieties I wanted to add to the collection. So I began interplanting with varieties on more dwarfing rootstocks between the rows, which is why there are, at the time of writing, 130 trees in slightly over an acre. The mixture of rootstocks, which in the long run will produce different sized trees, has parallels with forest gardening: canopy, sub-canopy, utilising different levels, etc., and will over time, I hope, make for a highly attractive orchard.

By year five, the orchard was almost complete and well on its way, so the nature of the work changed to a yearly programme of orchard maintenance as we shall explore in Chapter 13.

My field orchard is an ongoing experiment, as is the garden. Of my 130 varieties, some will like 600ft in the hills on a north-west facing slope more than others; some will thrive and survive with vigour. Others will not like it but just about survive, whilst others may not survive at all. All of which is fine; a form of natural selection, a reaction to the elements, to place, location, altitude, soil, drainage and the rest. In future times I may have a few less or, if I maintain a collection, it will invariably be a slightly different mix to that which I have now, at the time of writing. Again that is okay; all of this is an organic process, a learning curve. If nature is all about adaptation and suitability, then who am I to shove a load of fruit trees on a Shropshire hillside in an assumption they will all do my bidding. Of course they won't; that very genetic diversity that underpins their place in the *Malus* genus means some will do well, others will not and with an element of Darwinian selection playing out under my nose. After all, it may be my obsession, but it is only an orchard.

THE INTER ORCHARD – THE OPPORTUNITY FOR SO MUCH MORE

This is the point at which elements of the garden orchard and the field orchard potentially merge – an assessment of the spaces in between the trees and utilising that space to obtain other additional yields. The notion of multiple yields of production is a key aspect of permaculture and one worth considering at all stages and every opportunity.

Traditionally, as we saw in Chapter 1, we have perhaps come to see orchards as a group or line of trees with a grass sward beneath and perhaps a few sheep grazing, making them a dual land use system. But with good design and some creative thinking, why not a triple or multiple land use system?

In both the garden and field orchards, I have tried to recreate the Worcestershire tradition of daffodils beneath fruit trees

Some widely spaced orchards enabled a hay cut to be taken from the wide grass strips between them and during WWII, with the whole 'Dig for Victory' push, some orchards had rows of vegetables cultivated between the trees, but there are also examples from peace time of a more integrated approach to orchard production. In the Vale of Evesham in Worcestershire, there was a system of rows of currant bushes grown in the space between the rows of trees: a dual cropping system and dual income with the currant crop ready in July and the plum or apple crop two or three months later. Currants, being derived from a forest ancestry, are semi-shade tolerant. Some orchards also had rows of daffodils grown between the trees, giving the grower a cash income in February and March to boost the fruit income that came in September and October, a great additional income to aid the very uneven cash flow in farm businesses.

Daffodils, currants and plums are examples of three possible yields, but with some permaculture type designs, a whole raft of possibilities open up. Climbers up and through trees, fruit and/or vegetables in between, integrating the wider garden of fruit and vegetables amongst the trees of top fruit. There is perhaps a fine line between where an integrated orchard veers into being a forest garden, and we could spend a lot of time subdividing forest garden categorisation as opposed to orchards with extras. But we will stick to the premise that an orchard has a primary structure of a number of fruit trees. It is what we do with the space in between that is different here. In a limited space scenario, every square foot counts, so why not grow in the gaps in between?

This approach makes a great deal of sense and begins to introduce concepts like companion planting, layers of production, integrated pest control and pollination benefits. As the world population continues to grow and space is ever more at a premium, and food production issues become central, then it

Above: Comfrey makes for a great chop and chuck mulch as it mines minerals from deep down in the soil that can then be recycled by young trees

Right: Raspberries are essentially woodland plants so can grow in slightly shaded conditions beneath and between trees

could be seen as a decadent waste not to use those spaces in between. So the concept of the inter orchard is born of the garden and the orchard becoming one. Fruit trees do of course cast shade and that is a factor to consider, but not shade on a par with a forest or dense woodland, so with good design, careful planning and an understanding of the shade tolerance of different plant species, there is no reason not to use the inter orchard area to its full potential. Plants like currants, gooseberries and raspberries are all essentially woodland plants by ancestry and so will not mind a degree of light shade; again it is about the right plants in the right place and good design.

Traditionally, in the kitchen gardens of stately homes and large houses, fruit trees were not always a feature and were confined to a separate orchard away from the vegetables. So, perhaps the cottage gardens of the less well-off folk are a more fertile ground for evidence of orchard integration. Cottage gardens were, by their very nature, space limited, so a few trees in amongst the flowers and the vegetables were far more common. The cottage garden concept has other benefits, such as the notion that mixing crops together hides them to

In the garden pumpkins
weave between stepover
apples

some degree, and confuses pests and predators. Certainly a good mixture of fruit, flowers and vegetables all growing together in one place will draw in a greater range of insects, which will hopefully create a more natural pests-predator-prey dynamic on the plot.

At the garden scale, the possibilities are perhaps more obvious, doing reasonably conventional vegetable gardening in amongst and beneath the trees, building in layers by having currant bushes beneath, ground cover of alpine strawberries and perhaps kiwi fruit growing up through the trees. In my experience, beans don't seem to do well climbing up fruit trees, too shaded perhaps.

On the field scale, pests like rabbits may be a greater factor and if, like me, you don't live at your field orchard, then vegetables and crops that need more regular attention are not really as suitable, so fruit bushes and flowers may be a better fit. I have currants, gooseberries, josta berries and raspberries in places amongst my field orchard and whilst much of it feeds the birds, there is usually sufficient to glean a harvest for ourselves, with little to no maintenance. These perennial bushes largely just do their own thing, bar the occasional prune.

I have also planted several rows of daffodils to recreate that Worcestershire tradition. They can be a valuable nectar source for early emerging bumblebees and generally brighten up the view at what can otherwise be a dull time of year.

The actual interplay between plants is another technique we can draw from forest gardening. As we shall see, young fruit trees hate competition from grass, which competes with them in the shallow root zone. Comfrey is a very deep rooted plant which mines minerals and nutrients from deep down in the soil and stores them in the plant, so if you plant comfrey beneath or close by your trees you can cut it and mulch around the base of young trees to recycle those nutrients and trace elements that the tree itself may not be able to mine whilst it is still young and shallow rooted. So, comfrey as a chop and chuck mulch is an easy and worthwhile practice to incorporate in garden or field orchards.

Station Road Permaculture Garden is largely a garden with a lot of fruit trees in amongst everything else, whereas my field orchard is rows of fruit trees more akin to what we may consider a traditional orchard, just with other elements integrated amongst those trees.

Along one boundary I planted a 'fedge' or 'food hedge', consisting of jostaberry, hazelnut, blackcurrant, gooseberry, and raspberry with medlar and pear trees as standards within it. This notion of dual usage is a much-overlooked element of design. It acts as a hedge but is also made up entirely of edible species. Today, we perhaps just see hedges as those things around field boundaries, but many of the hedges of old were productive and rural folk would have harvested them for berries, nuts, kindling and firewood. There is no reason why we cannot recreate that tradition today. So, when planting a hedge, think creatively about what else it might yield.

> "The fruit of apples do
> differ in greatness, form,
> colour and taste; some covered
> with a red skin, others yellow
> or green, varying indefinitely
> according to soil
> and climate."
> John Gerarde, 1597

CHAPTER 6

Finding your Perfect Combination

Having done your research and tried as wide a range of fruits as possible, you should begin to formulate a 'want' list. For home use, this may be an early variety or two and a succession of varieties that will keep, spreading your crop over as long a period as possible. Or for a business idea of fruit, juicing, ciders or vinegars, you may hone in on a few very specific varieties tailored to your requirements. Again this will be bespoke and geared to your plans and vision.

HOW TO SELECT THE RIGHT VARIETIES FOR YOUR NEEDS AND SPACE

It is always worth dealing with knowledgeable, specialist nurseries, and getting the right rootstock for your situation is most important. It is a difficult balance, as standard big old-fashioned trees will ultimately look wonderful, be great for wildlife, enhance the landscape and are the key element of a traditional orchard. However, they are ultimately difficult to pick and prune, requiring ladders, and can seem daunting to manage, whereas bush or semi-dwarfing trees that are far smaller can be largely dealt with from the ground or a couple of steps at most, but obviously will never have the grandeur of the traditional standard orchard.

What to go for is a decision only you can make. In some circumstances your site will dictate the choice for you. Standard trees are simply not appropriate for backyards and small plots. Equally, if you are considering an orchard with

Always refer to a rootstock chart to ensure you get the correct size of tree for your situation

livestock involved then bush trees are equally inappropriate there. There is no broad-brush right or wrong. Every orchard will be bespoke, as it should be, suited to the situation, owners' needs and preferences.

The understanding of tree size and the different rootstocks is essential when purchasing trees. As we have seen, fruit trees are grafted and the root system onto which they are grafted is what dictates the size and vigour of the tree, whilst the scion or top of the tree dictates the variety of fruit that will be produced.

Rootstocks were once just those seedling trees used as a vehicle to carry the desired grafted variety. But in the mid-twentieth century, rootstocks, as we shall see, were standardised, so now all trees grafted onto say *Malus* M26 are growing on genetically identical roots to every other apple on M26. Some do consider this slightly questionable, as a degree of diversity has been lost and with so many trees on one single genetic rootstock it could potentially leave them open to a future disease outbreak.

It is of course worth remembering that the development of rootstocks was driven by the vast commercial sector, where growers wanted thousands of trees that all grew the same way, with the same vigour (or lack of it), behaved in the same way and cropped at the same time. It was not driven for, or by, the small gardener or permaculturist. On smallholdings and for amateur growers it is perfectly feasible to use seedling rootstocks in the way past growers did. Plant a row of pips, and when a year or two old, use them as roots onto which your chosen varieties can be grafted. You will not however know what the vigour of those trees may be until they mature, but if size is not a major consideration and you have room, then experiment and give it a go. You will have greater genetic diversity than using modern commercial rootstock.

When contacting a nursery to source trees you must be clear on the varieties you want, and on what rootstock you require them. There may also be a choice between potted trees and bare rooted ones; the latter have been field grown and then dug up in the dormant season (November to March) with their roots exposed. These bare rooted trees are, in my view, far preferable; they are cheaper, easier to transport and do seem to establish better as they have to accept the soil type of their new environment and get on with it, as opposed to the potted tree which can be reticent to spread its roots from the cosy compost of its potted cultivation. Given a choice, my advice is always select bare rooted trees.

Today the internet allows us access to countless nurseries and catalogues so do your research and shop around to get the right trees on the right rootstocks for your site. Any nursery of worth will be happy to advise if you explain your situation and what you are looking for. (For information on forms to consider, see page 85.)

POLLINATION

Many garden books and television programmes trot out the adage that pollination is a big issue and you need two fruit trees planted next to each other that flower together and so allow cross-pollination. However, bees and other pollinating insects travel widely, up to a mile or two around the locality, so to suggest the need for a neighbouring tree a few feet away is an insult to insects being able to do what insects do. They will find other fruit trees in flower at the same time and move between them. If you live in say the Highlands of Scotland, where there are far less fruit trees, it may be more of an issue, but in the most suitable fruit growing areas of central and southern England it simply isn't an issue, so you have no need to worry about it.

Bees and other insects will find other trees in flower and do the pollination naturally © Steve Hughes

ROOTSTOCKS: AN EXPLANATION

Rootstocks are effectively rooted sticks cloned from the mother stock

Fruit trees as we have seen are grafted to get a perfect genetic replica of a variety, and as most fruit trees won't easily root from a hardwood cutting like other trees, such as willow and poplar, we have to supply them with a new set of roots; so apples on apple roots, pears on pear and plum on plum roots, etc.

In Victorian times, rootstocks were simply grown from seeds, stones or pips and then had those chosen varieties grafted onto them. The problem with this method is there was no consistency in size, vigour or disease resistance because all the rootstocks were variable and random. So by the 1950s and driven by the commercial drive for uniformity and disease control, rootstocks were standardised. In a wide ranging series of trials, seedlings were grown and analysed not for their fruit but for the form and spread of their growth, their anchorage and for their vigour and disease immunity. From those trials a handful were selected and propagated to become the commercial rootstocks we use today. Each was selected for the size of tree and the form it produces; whatever variety you graft onto the roots, it is the roots themselves that largely dictate the eventual size and form of the tree. Hence we have standard rootstocks for large imposing trees, small dwarfing rootstocks for those small garden trees that never get too large, and in between we have a range of moderate or intermediate rootstocks so we can have the same variety but as trees of differing sizes.

Apples

The most common rootstocks you will encounter for apples are:

- M25: vigorous rootstock, used for traditional standard (large) trees
- MM111: vigorous stock, used for traditional standard (large) trees
- MM106: semi-dwarf / intermediate stock good for medium sized trees
- M26: semi-dwarf and the garden favourite for smaller trees
- M9: dwarf stock widely used for modern bush orchards, likely to need staking its whole life
- M27: very dwarfing stock, suitable for patio container apple trees.

It is slightly frustrating that the stocks still retain their original trial numbers and so can seem confusing to the amateur. So refer to a suitable chart and ask advice from a specialist nursery.

Pears

For pears the diversity of rootstock is somewhat less:

- *Pyrus communis*: effectively wild pear stock, very vigorous for large old style trees
- Quince 'A': semi-dwarfing for a moderate sized tree
- Quince 'C': dwarfing for small pear or quince trees.

Plums

- Brompton: big old-fashioned stock for full standard trees
- St Julian 'A': semi-vigorous plum stock for larger trees or standard plum orchards
- Pixy: dwarf plum stock for smaller bush trees.

Cherries

- Colt: semi-dwarf stock for moderate sized trees
- Gisella '5': a dwarf stock for small or bush trees.

Other modern rootstocks are becoming available in this ever changing field, so another reason to use reputable fruit tree nurseries to get the best advice to suit your situation.

Years later the union between rootstock and graft can still be clearly seen

Rootstocks are clones and are propagated by planting one which at a certain point is coppiced (cut down to the ground). It then produces multiple shoots which are earthed up as one might earth up potatoes. These new shoots then root into the soil (sawdust is sometimes used as an alternative) so when the earth (or sawdust) is brushed aside a year later, the young rooted shoots can be clipped from the mother stock, creating cloned copies. These mother stocks can go for years, becoming vast coppice stools yielding tens of stock each year or two. Just as the varieties we graft onto them are clones of their original seedling parent, the roots we graft them onto are clones, proving the alien nature of bought fruit trees, two cloned pieces spliced together in the middle!

To young trees an occasional soaking in dry spells can be the difference between life and death

DIVERSITY

If predictions on how the climate might change in the years ahead are true, and we can almost certainly expect some change, then we need to try and build resilience into our systems, and orchards are no different in this respect. So what can we do to live with a changing climate in terms of our own gardens and orchards?

Well, firstly diversity; don't put all your fruit in one basket. Literally grow a range of types of fruit crop, as well as a range of varieties within each group. By spreading the risk in this way you are far more likely to always get a crop of something. When the plums have a bad year, the apples may have a good one; when one or two varieties of apple have an off year, other varieties that blossom at different times, or can cope with different weather, will deliver.

Make the most of wet winters, and as they come to an end, get a heavy mulch in place (see Chapter 13), capping in all that winter rain and making it available to the trees for as long as possible into spring and early summer. If the climate

Large standard trees are incredibly difficult to work on

is going to be more extreme in terms of wet and dry, we need to try and combat those extended dry spells, by capitalising on moisture retention. With a really well mulched tree, it is amazing how even in an extended period of no rain, if you plunge your fingers deep into the mulch layer, you will find it moist beneath even when all around looks parched.

If you have any type of structure or roofed area in or near your orchard, harvest rainwater by installing butts or tanks to collect the runoff. Although watering orchard trees on any scale is impractical, if a real drought does hit, then certainly for trees in their first or second season, an occasional bucketful can be the difference between life and death. As with watering any tree, the occasional soaking is far preferable to regular small amounts. Tree roots need to be encouraged down and outwards; regular small amounts of water on the surface do not encourage this at all, so an occasional thorough soaking is definitely best practice when it comes to tree watering.

"On the motionless
branches of some trees,
autumn berries hung like
clusters of coral beads, as in
those fabled orchards where
fruits were Jewels."

Charles Dickens

CHAPTER 7

Going Beyond Apples

PEARS

This book thus far has been rather heavily orientated toward apples, partly due to my interest in them but also as they are the most studied, recorded, researched and the most widely grown of all the top fruit here in the UK and possibly in most temperate climates. Their cousin the pear is a fruit that has fallen so far from its pinnacle as to be almost entirely overlooked today. It is seldom seen or widely eaten, confined to perhaps one or two varieties in the supermarket or, worse still, canned in sickly sweet juices as to lose any concept of what a pear should truly taste like.

The pear was, however, once held in possibly greater esteem than any other orchard fruit. Until comparatively recently the number of pear varieties in cultivation greatly exceeded those of apples. In 1629 in his *Paradisi in Sole Paradisus Terrestris*, John Parkinson lists 66 varieties of pear compared with only 58 of apple, along with the claim that "the variety of pears is as much or more than that of apples" while in 1628, the French writer Le Lectier lists 260 different pears. John Evelyn in his famous *Silva* of 1678 listed only 27 apples compared with 57 pears plus four perry pears. Leap forward a couple of centuries and Scott's Nursery of Merriott in Somerset listed in their 1870 catalogue 1,022 apples, 58 cider apples and a staggering 1,538 pear varieties, including 68 which ripened after Christmas and 55 that only ripen between February and May in their 1870 catalogue, giving us yet another insight into the value and knowledge of varieties that would keep in the pre-supermarket and refrigerated storage era.

In some ways the pear was perhaps the connoisseur's fruit, for there is far greater skill and judgement required in the cultivation, storage and eating of pears than there is of apples. There is an old joke that there are only 10

The fruit of 'Worcester Black' pear, which keeps right through winter, will hang on the tree long after the leaves have fallen

minutes in the life of a pear when it is perfectly ripe, and that the skill lies in knowing when those 10 minutes are. The majority of pear varieties are not eaten from the tree but are picked underripe and then mature and ripen in storage, hence the skill is in their preservation. They can be kept in a cool environment for many weeks, whereas in a warmer one they will ripen much faster. In the heyday of pears and grand houses, gardeners and domestic staff, the pears would have been stored in a cellar or outhouse and brought in for a few days or a week prior to being eaten. Therein lay the cook or housekeeper's skill in judging that perfect ripeness.

Pear wood is hard and close-grained and in the past was used as a specialised timber for fine furniture and making the blocks for woodcuts. The production of pear timber may have given rise to

'Beurre Hardy', a pear once held in great esteem; judging its perfect ripeness takes real skill

the likely misconstrued saying, 'Plant pears for your heirs', which some have taken to infer that you will wait a long time to see any fruit when planting a young tree. It is my belief that this saying relates to the timber; by planting pear trees you are leaving your heirs something of value and worth way into the future. Many varieties, especially on modern rootstocks, will fruit well within four or five years, so seemingly proving that the saying does not relate to their fruiting characteristic at all.

Pears fall roughly into four groups:

Dessert pears

The pear equivalent of the eating apple and almost exclusively what you find in the shops today. Only three or four varieties dominate the market, especially 'Conference', 'Comice' and occasionally 'William'. They are picked and sold unripe, which means judging their ripeness is still necessary; many pears get wasted as today's consumer cannot gauge that level of ripeness correctly, either eating them when too hard or finding the pear has already rotted from the middle outwards.

'Doyenne du Comice' is considered by many to be the finest of dessert pears

Culinary pears

The pear equivalent of the cooking apple, these usually large, hard pears never ripen to the point where they can be eaten raw, and so have to be cooked. Often referred to as Warden

'Catillac' is one of the best keeping and culinary pears

Pears, they were once highly prized and were a regular feature of the Tudor recipe book, being baked, stewed and made into hearty pies. The oldest cookbook in the English language, the *Forme of Cury*, from around 1390, makes reference to cooked pears, and certainly by Tudor times they were a much valued ingredient in the kitchen. For centuries, most fruit was consumed cooked, the concept of eating it raw being a rather modern phenomenon. People of the past were suspicious of raw fruit, for they experienced what too much of it could do to the stomach and resulting bowel movements. They equally saw people die horribly from diarrhoea and sickness and with no medical understanding of why these things happened, they very literally feared that fresh fruit could be the death of them.

Stunning perry pear trees were a feature of the landscape of Gloucestershire and can live for 200 years plus

Perry pears

The pear equivalent of the cider apple. Almost all the varieties of true perry pear are inedible. To bite into many is to experience such astringency that it feels as though your mouth is being dried out. However, when correctly processed, they produce a juice that can be fermented into perry, an alcoholic beverage that some connoisseurs consider among the finest of all drinks. Perry has experienced something of a resurgence in the last 20 years, with craft and artisan perry being made and enjoyed in a way it had not been for perhaps a hundred years or more. In Napoleonic times, when drinking imported wine was considered unpatriotic, perry was the drink of choice of the ruling elite.

'Blakeney Red' a classic Gloucestershire perry pear

Historically, perry pear cultivation in the UK is unique to the Three Counties, Herefordshire, Gloucestershire and Worcestershire, and occurs nowhere else in the country. A culture of perry making does occur in other parts of Europe; it tended to use a more general mix of pears rather than the specific perry varieties synonymous with the Three Counties. The trees themselves can live for up to 250 years and where old examples survive, they make stunning features in the landscape, especially in blossom time. The county of Gloucestershire probably has more surviving ancient perry trees than anywhere else.

Harvest pears

These are the exception in the pear family, being small pears that are fit to eat straight from the tree and are earlier to ripen than most of the other types. They get their name from the fact they would be plucked from the tree and eaten fresh by labourers and rural folk working the land at harvest time.

The flavour of a perfectly ripe pear is sublime, and like nothing else, and it is tragic that they have fallen so far from favour, that so many people have never enjoyed this exquisite experience. So if planning an orchard, I would urge anyone to include a pear tree or three. I grow 18 varieties including four of those once renowned Warden types, the heavy-weight cookers of the pear world.

The little harvest pear, perhaps a close kin to its wild ancestor

A PEAR OF CONTRASTING STORIES

Two pears that span the history and popularity of the fruit and illustrate its fortunes are 'Conference' and 'Worcester Black'.

'Conference' is the most widely grown and eaten pear in the UK today, as well as being grown widely across the continent. 'Conference' was raised by the once renowned Rivers Nursery of Sawbridgeworth in Hertfordshire, from an open-pollinated Leon Leclerc de Laval (open pollinated meaning its other parent is unknown). It was exhibited at the National British Pear Conference of 1885, hence the name. It is well suited to the British climate and is a heavy, reliable cropper. The trade likes it because it keeps well and does not bruise as easily as other varieties do. As a variety to grow, it has this unusual characteristic of still being able to produce good fruit even when not pollinated, a trait known as parthenocarpic fruiting, hence its suitability to the vagaries of the English weather, which can play havoc with pollination. Its long pyriform shape makes it distinctive and a familiar sight on the supermarket shelves.

From the most common of pears to one of the least common: the 'Worcester Black' or 'Black Worcester' is synonymous with the city of its name and appears as an emblem on crests, brands and logos across the city. The story goes that Elizabeth I came to the city and commented on the fine pears she observed during her visit, and ever since it has been the emblem of both city and county. It is one of those big, hard cooking pears that will keep all through the winter and so would have been much valued in Tudor times. Its origins are completely unknown but it is an ancient variety, possibly the oldest pear variety still in cultivation. It is seldom found today and rarely eaten, which is a shame because it has much to offer in a range of recipes, a reflection on the decline in cooking fruit and the decline in the eating of hearty puddings. As a tree it is incredibly hardy, disease resistant and therefore easy to grow. The remaining ancient trees sit there in the landscape largely unnoticed, cropping away year after year, producing pears that almost no one ever uses.

Far right: 'Conference', the standard supermarket pear

Right: The 'Worcester Black' Pear is possibly the most ancient variety still in cultivation

THE PLUM

If apple, pear and plum are the holy trinity of top fruit grown here in the UK, then the plum is the most ephemeral of the three, perhaps trickier to grow than the other two and fleeting in its season. The trees are far shorter lived and far more susceptible to disease, making them harder to cultivate and crop successfully. Yet if you can get it right, picking your own plums warmed by the late summer sunshine is beyond comparison to anything you will ever see in a plastic punnet on a supermarket shelf. They equally fall into several categories:

Dessert plums

Almost the entirety of modern commercial production, the *Prunus* equivalents of eating apples, plums that can be eaten as fresh, sweet, flavoursome fruit. Even so, shelf-life is short and most will not last long without considerable degradation of flesh and taste. So, very much something to be enjoyed in peak season and then gone again for another year.

Culinary plums

Historically, the lack of eating fresh fruit, together with their inability to keep, meant the majority of plums were processed in some way, either by cooking, be it stewing, tarts and pies, or they were canned, jarred and jammed in huge quantities. In Worcestershire's Vale of Evesham, the plum industry existed on an extraordinary scale for 100 years or so, driving an entire local economy. Plums do lend themselves well to a wide range of processing; plum jam in my view is as fine as any jam, with exquisite flavour, and a good plum tart also takes some beating.

The 'Yellow Egg' or 'Pershore Yellow' was once widely grown for the jam, canning and bottling industry

Damsons, gages, bullaces, sloes and wildings

As well as the larger dessert and culinary plums, there are a range of smaller, less glamorous members of the family, with various names, be it bullace, damascene, mirabelle, damson and gage, they are basically all small plums often found in hedgerows, roadside verges and scrub or waste ground. In less affluent times they were enthusiastically harvested for stewing, jam and even the making of fruit gins. Today most folk don't bother with them and they are left to the local wildlife, but look around your locality and you'll likely find an example or two when out on a late summer walk.

Dessert plums are almost the
entirety of what is found in the
shops today

The 'Shropshire Prune' is one of the most
common damsons still found today

Shropshire Prune Damson

Unlike apples, many plums, and
certainly damsons and bullaces come
fairly true to their parents, that is to
say, grown from stones you pretty
much get the same or very similar to
the one it came from, so historically
many may have been grown from
stones rather than always being
grafted.

There are a wide range of gages, green and yellow; all
are basically small plums

Top: The huge majestic cherry trees belong to a past era

Above: The red cherry is the standard image we all have but they can be white or black

Right: If you can stop the birds getting them first, there is magic in a bowl of homegrown cherries. These are variety 'Stella'.

OTHER MORE UNUSUAL TOP FRUIT

Apples, pears and plums may make up the vast majority of our idea of a fruit tree for a garden or orchard, but with a little more imagination and an interest in history, there are rarer, more exotic and unusual fruits that can be considered.

Cherry

As ephemeral and fleeting as the plum, cherry production was once a considerable industry in Kent and Worcestershire, where vast orchards supplied the towns and cities with this most seasonal of fruits. In recent years, there has been a resurgence, with the area planted with cherries again, increasing after decades of decline. However, modern cherry growing is a world away from the majestic orchards of a century ago; today it is bush trees in polytunnels and a top end product with prices to match.

Cherries come in a range of colours and whilst most of us may think of the default cherry red, there are black and even white varieties that were grown in the past. In my part of the Midlands, some villages in the 1930s had a huge cherry industry worth tens of thousands of pounds in today's money, meaning vast effort was invested in the trees, their tending and the care and harvest of the crop, which could make or break the farm's economics in any given year.

If you wish to grow cherries in your garden or orchard, birds will be your biggest challenge, for their love of your cherries will be greater than yours and they rise earlier in the morning, helping themselves to your precious crop.

Quince

A relative of the pear, so close in fact they can be grafted onto one another with many modern pear trees being grown on quince rootstocks. As a fruit in themselves, they have a long history and a heyday perhaps in Tudor times when they were much prized. Generally an easy tree to grow, being hardy, and requiring little maintenance, with fruit left largely untouched by the birds. Added to apples or pears in a pie or tart, at a one to five or so ratio, it is considered by many to enhance flavour.

I like them simmered on the wood burner in a tiny amount of water with a spoonful of brown sugar until just soft. Then they have an exquisite, almost perfumed taste unlike anything else.

The quince, a historic fruit from Tudor and earlier times

The medlar is a strange and mysterious looking fruit that most people don't know how to use

Medlar

Another ancient fruit, almost entirely forgotten today, but possibly going back centuries, is again easy to grow but the trick lies in understanding the fruit and how to process it. They are productive and do well in the UK climate, but do not ripen on the tree to eating softness, so have to be 'bletted', that is, picked and then set aside and left to mature to a point of almost rotting. The flavour is somewhat an acquired taste. Historically they were eaten as a fruit as well as processed for stewing, jams and jellies.

Apricot

Until relatively recently, the apricot would have been considered just too exotic for the cool rainy British climate. A combination of less severe winters and the introduction of more temperate suitable varieties means apricots are now achievable in much of central and southern UK. Their prevalence for early blossom (flowering) is their biggest downfall, making them prone to frost damage and so ruination of the potential crop. If you can grow them in a sheltered spot, or keep the frost off with fleece or blanket, a good crop can be yours. I grow an American variety ('Golden Cott') that crops well at 600ft in the Shropshire hills.

They can be eaten fresh, dehydrated as dried apricots, or cooked, making excellent jam, as well as being wonderful in tarts and crumbles.

Peaches and nectarines

Another fruit that would seem far too exotic for our temperamental climate, yet in the right spot and in the southern half of our country you can get positive results. In other parts and with them now available on very dwarfing rootstocks,

Top: Even at 600ft in the Shropshire hills, apricots are possible on a sheltered walls

Left: With dwarf rootstocks, peaches can be cultivated in pots and moved into a greenhouse for the vulnerable flowering or fruiting periods

Right: Although considered more of a Mediterranean fruit, figs can crop well in UK conditions

they can be grown in containers and so put in greenhouses or conservatories for winter and when the blossom is setting (like the apricot, their early flowering nature can be their downfall).

Figs

Again a fruit we might not immediately associate with the UK, seemingly more at home in North Africa or the Middle East. Yet in the right position a really good crop can be had even in the Midlands. Normally grown against a wall rather than as a free standing tree, there are now a range of varieties available to grow here, but 'Brown Turkey' is the most common. I have one grown against a north-west facing wall that crops heavily in most years, whilst the perceived wisdom is that they need a south-facing sunny location.

Vines

If one fruit rivals the apple for its long and inseparable association with human history it is surely the grape, and again we may immediately think of the vineyards of France and the Mediterranean. Yet now grape growing and again good wine making is quite achievable here. The Romans had vineyards as far north as Wroxeter in Shropshire and today the West Midlands hosts award-winning vineyards. So on a garden scale, a couple of vines correctly established and maintained can yield a good crop of homegrown grapes for eating or fermenting.

My grapes are grown in a cold greenhouse, but good grapes are possible in central UK

Mulberries

This ancient and historic fruit resembles a dark raspberry yet it grows on a tree rather than a cane. They can become picturesque and gnarled in old age as well as being long lived and sprawling. Some ancient specimens from the 1600s and 1700s have toppled over only to carry on growing in a prostrate form. The biggest challenge to getting a crop is that the birds love mulberries even more than you do and as they become large trees, it makes them difficult to net, so the birds often win. A great tree to plant and nurture but perhaps don't hang too many hopes on abundant cropping. If you are lucky enough to harvest some, they make wonderful jam and coulis to accompany puddings.

Citrus

Definitely not a fruit that springs to mind when considering growing in the UK climate but as we have seen, the Romans did once cultivate citrus here, albeit in a rather different climate and probably in containers. So today they can really only be pot grown and taken undercover in the winter. Lemons are the most common, and are slightly more hardy than oranges and limes, but realistically anything below -2 or -3°C (28 or 26°F) and they are in trouble. Various stately homes had elaborate orangeries and invested huge degrees of care and considerable cost into their citrus trees, wheeling them outside in summer and back inside for winter. We have successfully grown lemon variety 'Four Seasons', and have been able to sit in the garden with gin and tonics with slices of

Lemon variety 'Four Seasons' cropping in Shropshire; it is more of a novelty than a serious crop in my part of the country

Almost unknown today, 'chequers' were once harvested from service trees

Shropshire lemons. But realistically growing citrus in the Midlands is a novelty and the cost of the trees and their care mean they are probably ridiculously expensive lemons. I do have a friend with an unheated conservatory who seems to be able to produce vast crops of lemons on a single tree in a large pot, much to my envy.

Wild service tree

A member of the *Sorbus* family being related to rowan and whitebeam, it produces berries known as 'chequers', which were once harvested and eaten. Stories of them being dried and given to children as sweets go back to Tudor times. Said to taste of dates when almost overripe, they were also brewed into an alcoholic drink, presumed to be why there are pubs called 'The Chequers'. So whilst perhaps not an orchard tree in the traditional sense, it can make an interesting addition to a hedgerow or something highly unusual in any planting scheme.

When designing any planting scheme, it is hard to over-emphasise the value of diversity. Plant a varied range of types and varieties and something will always produce for you whilst something else may have a year off. Diversity really is king.

"If you are going to add
fruit trees – and every garden
should have at least one –
think carefully where
to put them."

Alys Fowler,
The Edible Garden, 2010

CHAPTER 8

Sourcing your Trees

Having decided upon the nature of your new orchard and planned what you want to grow, be it apples, pears, plums, cherries or apricots, it then comes down to which of the multitude of varieties do you select for your plot? Fruit trees are a long-term investment and so need to be done correctly. To spend several years nurturing a tree that ultimately crops fruit that you find disappointing or unpalatable is an expensive mistake in time, labour and money.

I have lost count of the number of times I have been asked, "What do you suggest I plant?" to which I always reply, "I can't really answer that as I don't know what you like!" Whatever the fruit, everyone has their favourites; individual palates are highly varied and just as with wine or beer, the choice is vast and individual preferences equally wide. Sticking with apples, there are those two thousand plus British varieties. So palates aside, advising is impossible. Find ones that you like by trying as many as you can at farm shops, on friends' trees, at Apple Day events and markets. Seek out and try as wide a range as possible. Then when you discover those that satiate your taste buds, go and find that variety as a tree from a specialist nursery.

WHERE TO SOURCE

It is always worth using those specialist fruit tree nurseries as they will give good advice and supply clean, healthy stock. As much as we all like a bargain, cut-price trees from supermarkets and high street stores are, in my view, a false economy. Often imported from abroad, having been bare rooted and out of the ground for an unknown time, they often have highly limited information even to the extent of what rootstock they are on. Much better to support a local nursery, buy a British-grown tree and get the right advice with it.

My personal anecdote in support of this is that I once bought a discounted, end of season pear tree for £3 from a high street bargain store, which was labelled as 'Conference'. For two years it hardly grew at all but did eventually get going and then one year produced a crop of squat red pears – it wasn't even a 'Conference'. Lesson learned: far better to pay a fair price and go to a specialist.

As with all good permaculture, do your research and consider what you are likely to best use. In the case of apples that may be eaters, cookers, or even crab apples, but equally the same applies to pears, plums and cherries. Then plan for seasonality; using apples again as an example, those that ripen first (the first earlies) have very poor or no keeping qualities and need to be eaten within days of coming off the tree. Therefore, when selecting varieties, unless you have a use for a vast amount of early apples that don't keep, you will simply waste large numbers of them. So a few earlies to enjoy off the tree in August and early September are a joy, then have larger numbers of keepers, those apples that can be stored and eaten over many weeks and months into and through winter.

The objective for the home fruit crop is to make that crop last as long as possible, to be able to eat homegrown apples for the maximum number of days a year. Those really early ones like 'Gladstone' and 'Beauty of Bath' can be ready in July in a good year, followed by other earlies like 'Irish Peach', 'Discovery' and 'George Cave'. Then onto the second earlies like 'Worcester Pearmain', 'Devonshire Quarrandon', 'Epicure' and 'Tydeman's Early'. It is possible to plan a succession of ripening times, so as one variety finishes, the next is ready. When growing smaller numbers, such as on a cordon system, it is possible to plant a row of varieties that ripen at specific times and in doing so you can literally eat your way down the row. As one cordon tree finishes, the next one is reaching peak ripening. Like so much of permaculture, doing your research and making a plan is key, working it out in a logical and studied manner rather than rushing into impulse purchases and plantings.

In the last 30 years or so there has been a renewed interest in local varieties. For whilst we are likely all familiar, in the case of apples, with the mainstream favourites of 'Gala', 'Braeburn', 'Granny Smith', 'Golden Delicious' and 'Pink Lady', it seems tragic that of those 2000 British varieties, we are lucky if we see half a dozen in the shops. So when it comes to growing an apple, certainly don't plant something that is available in every shop; plant and cherish one of those varieties you can't buy anywhere.

Back in the 1990s, the wonderful little charity Common Ground kicked off the resurgence of local varieties and among their celebrations of 'the local' they produced the apple map of the UK in which they researched and listed the varieties of apple associated with each county, bringing to light many almost lost or forgotten fruits. Their work spawned a wealth of interest in local varieties and led to regional fruit groups and projects, apple detectives rediscovering the varieties of their areas (see Chapter 16).

There is a good logic to planting native varieties, those trees that originated and evolved in the locality. Whilst 'Gala' and 'Braeburn' originate from New

Zealand, 'Granny Smith' and 'Pink Lady' from Australia and 'Golden Delicious' from the United States (yes the States, not, as most people imagine, France), your local variety sprang from the earth of your own county and as such should be more suited to the soils, microclimate and pests and diseases of that area.

Put in the research and find out what the local varieties of your county, district or village were. There are now a range of local nurseries who propagate and sell local or heritage varieties. (See the useful resources section at the end of this book.) More and more people are now interested in that which is locally distinct, where their food comes from, how it was produced and in tasting, trying and ultimately eating that which is local.

Today with refrigerated storage and global distribution chains the concept of fruit as a seasonal crop is largely lost, when we can purchase the same Gala apple 365 days a year. But rewind 70 or 80 years and fruit of all sorts, including apples, was very much a seasonal crop. Hence knowledge and an understanding of that seasonality and keeping qualities was key. The first apples of the season would in a good year have been ready in late July; they would be followed by the second earlies ready in late August and early September and so on until the last keeping varieties came off the tree in perhaps November, and then were laid down as winter stores which, with some skill and luck, may have lasted through to the following Easter. After the last of those stores, probably slightly shrivelled by the end, there would have been no apples again until the earlies in late July or August. So whilst we still to some extent celebrate the seasonality of asparagus and runner beans, we have lost sight of the notion of apples, pears and plums as a seasonal crop. If as a permaculture orchardist you do not want to engage with the global fruit industry and its vast carbon footprint, then you need to think and design like the orchardists of 70+ years ago. Select a range of varieties that will give you as long a season as possible, plan it well and enjoy fruit as the seasonal crop it really is.

ORDERING TREES

- Having made your choices and integrated them into your plan (Chapter 9) you need to be clear in placing your order
- Ideally always use a specialist tree nursery
- State what varieties you are selecting
- Be absolutely clear that they will be on the right rootstock for your situation: M25, MM106 etc.
- Are you ordering bare root or potted trees?

Fifteen different seedling trees produce 15 vastly varied forms of apple

SEEDLINGS

Whilst we can buy this amazing array of varieties from specialist fruit tree nurseries, we can, if space allows, equally grow trees from pips, stones or seeds. They will, as we have seen, not be known or recognisable varieties but they will be unique, very genetically diverse and of varying vigour, health and disease resistance. But if you are so inclined and have the space then perhaps give it a go.

Six years ago I planted a range of pips in root trainers and waited to see what came up. I then grew on those that did come up in pots before planting out in our field. Six years on almost all have cropped and the range of forms, sizes and characteristics is considerable as can be seen from the photograph above, proving the extreme variation in the *Malus* gene pool.

None of them are especially noteworthy in terms of amazing flavour but a few interesting characteristics emerged: one hung on the tree until almost Christmas, another dropped suddenly before September was out. So a large crop of seedling fruit may not have an obvious use as dessert or culinary fruit but mine were perfectly usable as juicing fruit in a mixed pressing. (See Chapter 12.)

Observing seedling trees is not all about fruit; the vigour and form of the trees themselves is as diverse and variable as the fruit. Some can be incredibly vigorous, some have very upright form. Some are spiny with almost thorn-like protuberances. Some have rock solid anchorage and others seem less well rooted in the earth, some seem in fine fettle health-wise, others sickly and disease prone. Just like people, all seedling trees will be different. So whilst certainly not for those keen to cultivate a few trees for maximum harvest, if you have land, the space and the inclination, give a few seedling trees a go. It is truly a one in a million shot but you might produce something truly amazing.

One of my seedling trees that shows promise as a decent cooking apple

GRAFTING

The opposite of those seedling trees, as we have mentioned, is the grafted tree. Grafting is that almost miraculous technique that never ceases to fill me with a schoolboy sense of wonder, in that you can chop a piece off one plant and stick it onto another and away it grows. We know that the Greeks and the Romans grafted, so it's an ancient practice and one is left to ponder who it was that had the first light bulb moment of thinking 'I wonder if I cut a piece of this tree and splice it onto that tree'. Perhaps they were inspired by observations in nature where occasionally plants of the same species fuse together naturally. However it first came about, it has been the mainstay of fruit tree propagation down the centuries and as we have seen it is the only way to ensure an exact genetic copy.

While probably not for the novice orchardist, once you get more established and confident, it is perhaps a natural progression to want to graft one's own trees and it certainly makes for a huge financial saving. Rootstocks can be purchased from specialist nurseries and it is then a case of sourcing a scion (a cutting) from the variety you wish to propagate. Friends' trees are a good source; ideally you want a pencil length and thickness piece of last season's growth. I always graft in the first two weeks of March as it is just before the growing season and means the newly grafted stock don't have to sit dormant for too long until the magic of spring kicks in again and new growth commences.

Whip and tongue grafting is the most common technique for the amateur and involves making a spliced cut on the rootstock while you then make a corresponding splice on the scion with the aim of marrying up the cambium layer. Yes, remember those school biology lessons? Grafting is all about the cambium layer, getting that living part of the bark to marry up, scar, heal over and fuse so that the rootstock adopts and feeds the scion with sap, a union that will last for the rest of the tree's life. This can be done as bench grafting, where

PROPAGATION GRAFTING

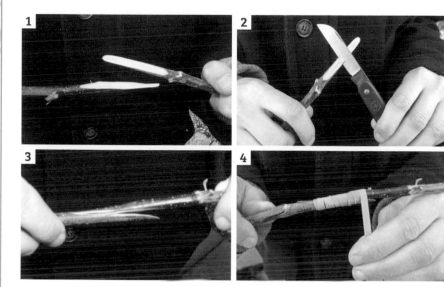

1. Both the rootstock and the scion have equal splices to expose maximum cambium

2. A small counter nick is made

3. Enabling the two parts to be pushed together forming a join

4. The join is then firmly bound with a grafting band

5. Young grafts are just starting to grow in pots in the greenhouse

6. By late May the newly made tree is well on its way

the rootstocks are bare rooted and hence easier to handle, or it can be done in the field where the rootstock remains firmly in the ground. The former is far more conducive to the human back. Bending over all day grafting between your ankles out in the field is a less than enjoyable way to spend a day. Ideally the rootstock and the scion (cutting) should be that pencil thickness, making for a compatible union.

There are multiple other techniques of grafting such as chip budding, T budding, cleft grafting etc., but all are variations on that theme of cambium fusing to form a union.

CONVERSION

The other method that may interest all orchardists is that of tree conversion, taking an already reasonably established tree and either changing the variety or adding further varieties to it. The aim is the same, to marry up that cambium layer, but the dimensions of the material are wildly different, requiring different techniques. Today in the age of bush orchards with short lives, where they are grubbed up after only a few years, the concept of orchard conversion is largely lost. In the past it was possible, if you had say an orchard of 'Russet' apples and the market for 'Russets' faded, you could cut back the tree's crown (branches) and overgraft with a load of scions of another variety and in two or three years' time those original 'Russet' trees would produce a crop of a completely different variety of apple. Being placed on an already established tree with its extensive root system, new scions can romp away and be cropping inside two or three seasons.

One beautiful old orchard in Worcestershire's Teme Valley has towering apple trees where the 85-year-old owner tells me virtually none of them have a trunk that is of the variety of apples produced in the crown because back in the 1940s and 50s his father had had many of them overgrafted, another example of the unnatural nature of fruit trees and how men and women down the centuries have manipulated nature in extreme ways for productive ends.

I have been in several of these ancient orchards where the apples being produced in the crown of the tree are not of the variety of their mighty trunks. Another example is where an unnoticed shoot may get away and over time become a productive branch and so you may notice two different apples or pears being produced on the same tree, one being the variety, the other usually the rootstock cropping. In the case of apples and pears, the fruit of rootstock is usually small and unpalatable, a reversion to the *Malus* or *Pyrus* ancestry.

If you reach a point in your orchard journey when you have the confidence and interest to graft, give it a go; it is a great skill to master and there is huge satisfaction in creating your own trees.

CONVERSION

In 2011, I converted an apple tree of the variety 'Gladstone', grafting on three different varieties.

The young tree is headed back so all its branches are cut back to short stubs

The bark is then cut down its length and lifted to open it up rather like a shirt collar

A splice cut is then made on the scion and it is tucked under the bark

Multiple scions are put on, then bound with twine and sealed with a paste to prevent drying out and bacteria getting in

Not exactly pretty but the tree is now overgrafted and within a few weeks the new scions start to break bud

By early summer they are in leaf and the tree is now growing its new variety

By the end of its first growing season the tree has a brand new crown

Three years later the tree formerly 'Gladstone' is now cropping 'Scotch Bridgets'

Pitchers

As we have seen, virtually all cultivated fruit trees are grafted onto selected rootstocks which dictate the size and vigour of their growth, but there are a few varieties of apples that defy the general rules and for reasons unbeknown will fairly easily root from a cutting or torn off limb. Known as pitchers, for their ability to be pitched anywhere and take root, they would have once been prized, especially in remote rural areas or among the poorer folk, for you could simply obtain a tree by taking a piece off your neighbour's tree and plunging it in the ground. It is possible they were passed down from one generation to another so some could be very ancient in origin.

There are very few surviving varieties that have this characteristic but Ireland and Wales seem to have a disproportionate amount of them, possibly suggesting they either do better or are a product of the wetter climates of the west, or that these are simply the areas they have survived and not been as easily replaced by more modern varieties. They are believed to be incredibly hardy and adapt to cold temperatures, high rainfall and even salt laden winds. A clue to a variety being a pitcher is the formation of aerial roots, burr-like growths with small red rootlets visible; any piece removed with such a burr should easily root.

Some varieties do seem to have the ability to simply root from a cutting or torn off limb

These trees growing on their own roots are obviously more natural than a grafted tree and the vigour will be what it will be. Growing pitchers is an almost lost concept today, but for the amateur or permaculture grower perhaps one worth exploring.

> "In an orchard there
> should be enough to eat,
> enough to be stolen and
> enough to rot on the
> ground."
>
> James Boswell, 1807

CHAPTER 9

Making your Planting Plan

When working out your planting plan there are a number of other factors to consider that are more important than calculating it purely on a sheet of squared paper in a mathematical sense. The distances between your trees may be dictated by the rootstocks onto which your trees are grafted, and how you manage the grass or vegetation beneath the trees, especially if you are using a mower, a tractor or ride-on. (See Chapter 11.)

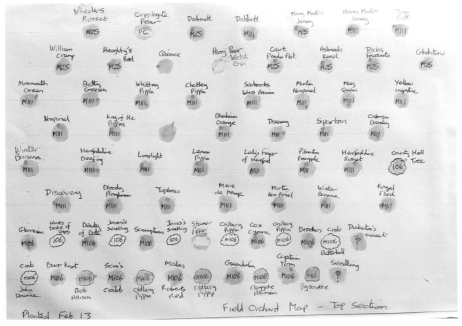

Make a plan that works for you and records what was planted where

ACCESS

In a garden situation, factors like sheds, greenhouses, paths and washing lines may all dictate which space can take a tree, and remember, trees will get a lot bigger than their initial planting size, so try to envisage them at full size and ask, "Will that size tree fit in this particular space 10, 20 or even 50 years from now?"

There is an element of subjectivity involved in design; some people like straight lines, others don't, so whether you plant your trees in rows or not will be a personal preference. For any form of mechanised grass cutting, straight lines do make for an easier, more efficient life. Prior to planting it is worth putting a cane or a stake in the spot where each tree is envisaged then have a look at it; try and envisage what that would look like as trees. Live with it for a day or two and if it doesn't quite look right, move the canes around, have another look and when you are finally happy with the way it looks, these are the points at which you will plant the trees.

ORIENTATION

A plan will be very personal; what one of us likes the look and feel of will be different to what another may like; straight lines, gentle curves, different heights, and that matter of what else you might wish to use the space for, will all be factors. But there are aspects we cannot change: elevation of plot, north or south orientation, neighbouring features, etc.

If planting a mix of different sized rootstocks, then an awareness of the north, south, east, west orientation of your plot will be helpful. In a planting scheme that will ultimately produce a mixture of sizes of trees, putting the shorter, more dwarfing trees on the southern side (northern side if you are in the southern hemisphere) will ensure they receive maximum sunshine. Placing the taller standard trees at the back or northern side of the plot thus makes the best use of the way sunlight falls across your site.

Shade is a factor to consider, for fruit is ripened by sunshine, so as a generalisation most fruit will not do as well in a shady spot. If you have walls, trees or hedges neighbouring your site, then plant well away from them, for the trees' quest for light will pull them towards the sun and can lead to angled trees with a disproportionate amount of growth on the sunnier side. Equally in the case of a south-facing aspect, walls can be a wonderful resource and for fans, espaliers and other trained forms, the passive solar gain and heat retention properties of a wall offer a great opportunity, creating a microclimate well suited to the more tender fruits like peaches and apricots, but apples, pears and plums will also thrive on a sunny wall. Visit the walled gardens of countless stately homes to see how the gardeners from the past understood this well.

Understanding the orientation and how the sun moves across your plot is key to maximising photosynthesis

It is sunshine that ripens the fruit, so as a general rule, fruit doesn't do as well in shade

Thinking about access paths and the way you will move around your site is also key, so plan gateways, paths and clearings. A large orchard does not need to be wall-to-wall trees; space them out, have open areas. Again, drawing out and marking it out with those canes will all help you visualise the form your orchard will take. Play around with the layout multiple times until you are really happy with it. It is going to be with you for a long time, so needs to be right.

TREES AT ALTITUDE

There is a fairly widely held notion that fruit trees don't do well above 600 or 700ft (182 or 213m) in the hills. Quite where this belief originated is unclear, for it is clearly not the case. In the uplands we can find many examples of orchards at height, old farm orchards at 1000ft or more. So fruit trees like many other plants and animals are far hardier and more adaptive than we give them credit for. Whilst walking the Offa's Dyke long distance path, on the Welsh / English border, a member of the Marcher Apple Network found a range of varieties thriving in ancient farm orchards well above 1000ft. So if planning an orchard at altitude, consider choice of varieties as some do better in higher exposed locations than others. Anchorage is key, because at height it can be windy; the reason my own orchard is largely on M111 rootstock is because it is renowned for good anchorage characteristics.

PROTECTION

Having invested time, money, effort and even love into planting your young tree(s) the worst possible outcome is that it is damaged or doesn't survive at all. There are a number of potential threats to your little sapling: rabbits, voles, deer, sheep, cattle, and even horses will all happily eat your tree, so an appropriate level of protection is necessary. A knowledge and assessment of your site will ascertain the likely threat. If rabbits are the only likely problem then a small spiral or tube, costing a few pence, is sufficient.

The larger the potential tree muncher, the more elaborate and expensive the protection becomes. There are a wide range of guards that can potentially be deployed, and knowing the threat is key, avoiding over-engineering with expensive guards when a simpler, smaller one would suffice.

In terms of domestic animals where orchards are grazed, trees are easier to protect from sheep than cattle or horses. A 4ft (1.2m) high protection weld mesh, mega mesh or post and rail, will keep sheep away as opposed to the vast reach and sheer bulk of cattle and horses that require much larger and sturdy guards. Ideally, if looking for a grazing mechanism for an orchard, sheep or geese are by far the best option.

Wildlife threats are a slightly different matter; muntjac deer are widespread and seldom seen so just because you haven't observed any, this by no means guarantees they are not about. Muntjac and rabbits are the most common threat to young trees and luckily, neither being particularly large means a relatively moderate and easily constructed guard will do the job.

One approach is to keep a careful eye on your trees and at the first sign of any damage be prepared to respond swiftly by fitting protection. Alternatively take a no-risk approach and fit guards on all trees from the start, but obviously there is a cost implication on this as the guards and stakes can cost multiple times that of the tree, not to mention the sheer effort and labour required in their construction.

Whether to stake young trees is a subject that divides opinion. Traditionally trees were always given a robust stake and lashed to it. However, more modern thinking speculates that allowing the tree to rock in the wind encourages a greater degree of rooting and anchorage, leading to firmer, more resilient trees. If a young tree is firmly secured to a solid stake for 10 years or so, could there be a danger that, when eventually removed, the tree is more likely to blow over as it has never needed to lock its roots as firmly into the earth as it was totally dependent on the stake? The science is not proven but if you can avoid a stake, it saves money and takes a more natural approach as trees in nature have to manage and get on with it. I try to avoid staking trees wherever I can.

DIFFERENT TYPES OF ORCHARD GUARD

1 Standard forestry tubes will suffice against hares and rabbits

2 Mega mesh is suitable against sheep, muntjac deer, rabbits and hares

3 Weld mesh available in various heights is strong and long lasting protection

4 Post, wire and rails are suitable against small breeds of cattle such as Dexters

5 Substantial cattle crates are needed against large breeds of cattle or horses

Shelter

My field orchard is very exposed and was created by fencing a one-acre corner out of an overgrazed 10-acre sheep field, so it already had established, if over-flayed, hedges on two sides and a newly erected stock wire fence on the other two sides. In that first season, we planted a mixed species native hedge on the inside of the two new stock netting fences, which I have subsequently allowed to grow up untrimmed. In seven years, it has grown a good 12ft (3.6m) tall and creates a good band of shelter on the northern and western sides of the orchard. I have also allowed the existing over-flayed hedges to regenerate by ceasing to have them cut, and so the orchard now has a far more enclosed and sheltered feel to it. Hedges allow some air flow through them and are therefore far better than any solid fence or barrier would be. The hedge also creates a habitat, a new opportunity for wildlife: within three years there were birds nesting in it, it is constantly worked by a flock of yellowhammers, and solitary wrens are seen ducking and darting through it. Pigeons nest in it and a sparrowhawk is sometimes seen cruising its length.

For exposed locations, shelterbelts, be they hedges or a line of some other species of tree, are hugely valuable in providing shelter, breaking the speed of the wind and protecting blossom and fruit, as well as the trees, from the worst of windy weather. In 2020, looking back over the orchard's development in the previous seven years, I honestly think planting the new mixed species hedges was key; the best investment in the project and the success of the fruit trees owes much to the shelter that the hedge provides at 600ft on a windy north-west facing slope.

It was not an expensive exercise with the hedge plants costing (at that time) around 25-30 pence each and planted at four per metre in a double staggered row. With the addition of plastic spirals to keep the rabbits at bay, again, at 25 pence a piece, establishment costs were under £5 per metre. A great investment in the overall establishment of the orchard.

Diversity is always to be recommended and the more varied the species mix, the more its value to wildlife is likely to be. I opt for a mix that is fifty percent hawthorn (*Crataegus monogyna*) and then ten percent each of five other species that may include hazel (*Corylus avellana*), crab apple (*Malus sylevestris*), guelder rose (*Viburnum opulus*), spindle (*Euonymous europeaus*) and wayfaring tree (*Viburnum lantana*). The two I would avoid are blackthorn (*Prunus spinosa*) for its invasive nature as it will sucker many feet out into the orchard, and dog rose (*Rosa canina*), because it is a thug and romps all over the hedge as well as possessing vicious thorns. A good mixed hedge, minus those two, is a wonderful shelterbelt and habitat.

The mixed species native hedge was arguably the best investment of the project, as it grew to give shelter to the orchard. 2014, 2018 and 2020 respectively.

FORMS TO CONSIDER

When it comes to the trees themselves, in the field situation it will be a choice of large trees or smaller semi-dwarfing. Whereas in the garden, the full range of trained forms give you options.

Stepovers

These were a Victorian creation to edge the kitchen garden in a decorative, yet productive way. Consisting of dwarf trees that have just two branches trained left and right, all the resulting annual vertical growth is cut back each summer to form spurs. Not only do they visually enhance the plot, but they can be incredibly productive with as many as 50 apples per stepover. They take their name from the fact that at only 12-18 inches from the ground they can literally be stepped over to access the plot beyond.

Despite their small size stepovers can be immensely productive

Cordons

Of all the trained forms, these are the easiest to create and maintain for the novice or less experienced grower, consisting of a young tree (often called a maiden) planted at a 50 to 60 degree angle, the belief being that the angle reduces vigour. All the side branches are annually trimmed back to four or five buds / leaves to form fruiting spurs along the length of the trunk, so effectively, a tree that fruits along the entirety of its length, whilst having no significant side branches. Maintenance consists of a five-minute prune, once or twice a year. For beginners, I always suggest cordons as the best entry level into the world of trained fruit.

Espaliers

This classic elegant trained form is often seen in the walled gardens of stately homes, but to train and create one is a long-term project requiring considerable skill and experience. Whilst they can be bought already started or created, they are not really one for a beginner. The principle is somewhat akin to that of a multi-layered stepover but nothing like as easy to master.

Trees over an arch

Fruit trees trained over an arch make for an attractive and productive feature; here a green cooker ('Catshead') on one side and a red eater ('Madresfield Court') on the other side are trained over a cheap metal arch from a gardening catalogue. Ultimately the steel frame may be removed, leaving a living productive doorway.

Dwarf or bush trees

These are ideally suited to small gardens. The traditional orchard tree has a clean trunk up to some predetermined height. But in a garden where there are no grazing animals, the trunk can be as short as you like, with branches coming from as low down as you like. Starting with a tree on a smaller rootstock, it will only get as large as you allow it to. Good pruning can shape and control a tree to whatever size you wish it to be. Several bush or dwarf trees make an ideal garden orchard, without ever needing a ladder or even a set of steps: easy to manage, maintain and pick from.

Standard trees

Not suited to any but the largest of gardens, for whilst the romantic notion of the old sprawling apple tree with a seat beneath, or hammock struck from its branches is an appealing one, they simply take up too much room in most domestic settings, and without knowledgeable pruning, have a tendency to run riot and end up dominating spaces and creating a problem. But if you do have a large plot, by all means, have a standard or two, but placement is key. When deciding where to plant, try to visualise what it will ultimately look like, and how much space it will occupy 20 years down the road.

Above: Cordons are the easiest trained form for the novice

Right: Apples over an arch

Below: Whilst espaliers look spectacular it takes real skill and time to create one

'Discovery' on M26 rootstock, a perfect small, domestic garden tree

Standard trees look lovely but need a lot of room

Varieties in the garden

We have 30 varieties of apple, five of pears and two of plums in our garden, and when planning your garden orchard, research and understanding seasonality is key. As mentioned earlier, you are unlikely to want a glut of any one thing, aside perhaps from a really good keeping variety. So select a range of varieties that ripen throughout the season with a few keepers, so that you can eat your way around the garden; soft fruit and plums can be frozen for out of season use. There is no substitute for that sense of expectation of the crop and enjoying it at the right time.

I planted my garden and its fruit trees before I had the field, and in hindsight may have done it differently; the nature of top fruit is well suited to an orchard being away from home as the season means harvesting can be done on occasional visits as opposed to vegetable and salad crops that need to be harvested and tended on a daily basis. At my field orchard I have 75 varieties of apple, six of pear and a smattering of damsons, plums, quince, medlar, greengage. Thus, were I starting with what I now have, there would be fewer fruit trees in the garden. For those with only a garden, obviously the space available and what you feel you are likely to need and use will influence your choices.

This tree in my front garden was originally a crab apple but now has six other varieties grafted onto it

Two scions added onto the family tree, their development and the same union years later

Family trees

One option for small gardens, allotments and backyards where space is at a premium is to have family trees, that is, a tree with more than one variety growing on it. Through the wonder that is grafting, it is possible to have multiple varieties growing on a single tree. Some nurseries and suppliers sell pre-grafted trees of three varieties but for those who have mastered the art of grafting, it is a simple task to add extra varieties onto existing trees. In my front garden I have a tree with seven different types of apple growing on it. Originally a crab apple bought from a garden centre, I added different eaters over the years so that each limb now produces a different apple; with the purple 'Spartans', the light green 'Limelight', the red of 'Kids Orange Red', the yellow of 'Herefordshire Russet' and the vivid crimson of 'Scrumptious', it is a multi-coloured spectacle of apples. Family trees represent a really good option where space is limited, enabling a far wider range of varieties to be represented than would be the case of single variety trees.

CHAPTER 10

Giving your Trees a Good Start

As we discussed in Chapter 4, there is more to an orchard than planting some trees and walking away. Success is based on tree care and maintenance of your trees. Giving your tree the best start in life and then some tender loving care will set it on its way to what will hopefully be decades of healthy productivity

TOOLS OF THE TRADE

It sounds like a horrible thing to say, but the advice I always give when delivering workshops and orchard training sessions is "buy yourself a good pair of secateurs and a good pruning saw and never loan them to anyone". It seems a selfish and unfriendly adage, but it is true. When it comes to tools, always buy the very best you can afford, and then look after them well, and that includes not loaning them to others; they will then generally last you a lifetime.

When it comes to making your spending choices, the range is vast, as is the associated price, and it does depend what your level of usage is likely to be. If you are simply looking after a couple of trees in a garden or on an allotment, using them a handful of times a year, then those tools at the cheaper end of the spectrum will suffice. If you are a borderline professional nurseryman or woman, then a top end pair of secateurs and the finest Japanese pruning saw will be a very worthwhile investment. So, you take your choice and pay your money. Good tools are a joy to work with; despite owning a good number of them, I have my go-to firm favourites that I will reach for time and time again.

As far as orchards go, the tools and equipment required are comparatively minimal, compared to many hobbies or professions. The basic requirement when it comes to maintenance is a kit consisting of a pair of secateurs, a good

The basic orchard toolkit is a good pair of secateurs, loppers and a pruning saw

pruning saw and a pair of loppers; perhaps a pen knife as well, but really a toolkit you can cart around the orchard in a bag or a barrow, with ease.

For planting and other activities like mulching, a good spade, a muck fork (a four pronged pitchfork), a wheelbarrow and perhaps a shovel are useful. What I like about orchard work is that it is largely hand tools, so peace and quiet can prevail while you work, and there is great satisfaction to be had from doing things by hand, as opposed to using noisy power tools.

My one concession to a fossil fuel guzzling, noisy machine, is a small petrol-driven chipper. Prunings, and what to do with them, is something that has changed over time. In the adage of old, a good bonfire and an obsession with tidiness and orchard hygiene prevailed. But to burn the branches arising from your pruning is not only an unnecessary contribution to climate change but is a waste of what could be a useful resource. So, by chipping the prunings back onto the orchard floor, we not only mulch and recycle nutrients and trace elements, but we are building soil health and fertility along the way.

Whilst I do chip a considerable amount of the prunings from my orchard as we will see later, there is also a good case for piles of logs, branches and sticks as home to wildlife who then, in turn, may fulfil useful pollination and pest control duties in the orchard.

Times change, as do views, and recycling and an appreciation of soil management means it is okay to chip and mulch, aside from extremely diseased material that is probably best removed from the orchard altogether. Such material I burn on my wood burner, thus utilising the heat to warm the home.

GROUND PREPARATION

The greatest struggle your young trees will encounter is compaction. Many soils right across the world are hard and compacted from years of mechanical tilling, the feet of countless grazing animals or just hard garden soil. So the best ground preparation is to loosen and aerate the planting area to give those young roots space to establish and expand. Whilst in the vegetable plot I am a keen advocate of no dig, tree planting holes are an exception, so dig over and loosen a hole considerably wider than the roots of the young tree.

It would seem conventional gardening wisdom is to enrich the planting hole with compost and even fertiliser but this can be counterproductive, for if the planting hole is a considerable contrast to the wider surrounding soil, there can be a reticence on the part of the roots to leave the cosy fertile confines of the planting hole and venture out into the, perhaps poorer, wider soil. This can be a particular issue with pot-grown trees where the roots have become accustomed to compost of the pot and in some cases become slightly pot bound. I have seen potted trees planted out that, two years later, can almost be lifted out of the planting hole again, the roots simply not having spread out.

So, over-enriching the planting hole is not good, as you are effectively creating a rich pot-like environment surrounded by contrasting, usually less rich, soil. However, one addition to be highly recommended is mycorrhizal root powder. These naturally occurring fungi aid root establishment. In a woodland it can be naturally occurring, but in former arable and pasture land or brown field sites it may be absent or in very low levels, so using an additive gives the tree an initial boost and is widely advocated by most nurseries as good planting practice. Widely available in garden centres, nurseries and online, mycorrhiza is a natural product, so usable by organic and conventional growers. When used in powder form, it is important to get direct contact on the roots, so by wetting the roots first, the powder sticks onto them prior to backfilling the hole. (See Chapter 5.)

PLANTING

When it comes to the actual hole, it should be bigger than the tree's established root system so no bending or cramming of the roots is required, and there is a school of thought that a square hole is preferable to a round one; like with the potted tree, there may be a tendency for the roots to get to the edge of a dug round hole and go round in circles as opposed to out into the wider soil. The thinking here is that the corners of a square hole prevent the roots going round in circles and force them to head out beyond the planting hole.

When placing your tree in the hole, depth is critical: the graft union must be above the height of the soil and trees planted too deep will not establish well. There is, with bare root trees, usually a discernible dirt line on the tree that indicates the depth at which it grew in the nursery and this should be replicated in the field situation. Fruit trees generally do not like sitting in wet and saturated ground, so on some sites the trees are planted on slightly raised mounds or hummocks to aid drainage and aeration of roots. Many old orchards

On wet sites, try planting on a slightly raised mound   A well dug and broken up hole is key to good planting

were planted on 'ridge and furrow', an old feature created by years of medieval ploughing whereby fields end up with a corrugated surface, and the fruit trees are invariably planted on the ridges to avoid sitting in the wet.

So having dug a big enough hole, inserted your young tree at the appropriate depth and sprinkled on the mycorrhizal root powder, backfill your hole firmly with the soil and, if possible, give it a thorough soaking, a can or bucketful per tree, as this settles the soil particles and air pockets and gives your tree the best possible start in life.

Whilst bare rooted trees can be planted any time from November to April it is preferable to get them in at the beginning of the season so they settle and get well watered in by the winter rains and before the ground gets too frozen and cold. If planted in March or April, a dry spring can cause water stress and high mortality rates as opposed to trees planted before the new year. A good best practice is, if possible, to get your trees in before Christmas.

WEED CONTROL

Young trees of any sort hate grass growing around their base as it competes with them for water and nutrients; it can vastly reduce their growth rates and even cause the tree's death in a dry spell. Anyone familiar with turf will know the matted, thick set of roots grass possesses, so by allowing grass and weeds to grow around the base of a young tree it is obvious that with its roots near the surface, it will have to compete. So keeping a clear area around the base of any young tree is key to successful establishment.

There are various options to achieve this. Mulch is the ideal: a good layer of organic material that suppresses weed growth, caps in the moisture and reduces the roots drying out. It will ultimately compost down and feed the tree. Fresh chippings from tree surgery are ideal, but old hay, straw, rotted farmyard manure, leaf mould and grass clippings are all good mulching materials. However, availability of such materials can be an issue and on a large scale the sheer volume required and labour necessary to move and spread it can be impractical.

In commercial orchards an application of herbicide is used to create a bare, scorched earth appearance beneath the trees that is unsightly, leaves the soil exposed and is of course of no habitat value at all. In my field orchard (see Chapter 5) I did at the outset apply a light spot spraying of glyphosate around each tree for the first two years in addition to some mulch. Glyphosate is now highly controversial and the ever emerging picture around the world and associated court cases related to it mean we should think very carefully about its use. I did what I did then but in light of more and more evidence, I am not sure I would do so now. I am left wondering whether we will one day view glyphosate in the way we now view DDT; sometimes what seemed a wonder at the time ends up having nightmarish repercussions.

Keeping that weed-free zone around the tree for the first five years will deliver far greater results in establishment and growth

A young orchard being planted with a thick mulch around each tree

An organic option is mulch mats or effectively a skirt around the base of the tree to prevent weed growth. This can be permeable like hessian or old carpet or it can be non-permeable like black polythene; either option can be very effective, but the one cautionary element is that on a field scale, voles can find living beneath mulch mats much to their liking and then in turn feed upon the bark and roots of your young tree. I am hesitant to use such mats for this reason and prefer woodchip mulch.

It has long been stated that using fresh woodchip or waste from tree surgery as mulch was a no-no because it supposedly stripped nitrogen out of the soil as it decayed. Recent research, however, has suggested that a thick layer of fresh green chip with a high level of leaf content hugely stimulates mycorrhizal activity as well as containing phenols and other valuable trace elements, so seemingly another old horticultural chestnut that belongs in the past.

Soil health and the micro life within it is a vast subject that we barely scratch the surface of and are only now beginning to fully comprehend through the regenerative agriculture movement. But it seems logical and common sense that a healthy, living soil will lead to healthy, vibrant trees, so rich mulches, feeding the soil and not poisoning it with chemicals is obvious. I am no expert but from my observations and current experience, five years of now being chemical free seems to be yielding a healthy orchard as I will explore in Chapter 14.

Whatever your chosen method and however you do it, DO control weed growth around your tree; it will pay off hugely every time.

> "In considering a
> Permaculture as a complete
> ecosystem, animals are
> essential to control vegetation
> and pests and to complete
> the basic nutrient cycle
> of a farm."
>
> Bill Mollison, 1991

CHAPTER 11

Dealing with What Lies Beneath

WHAT ARE YOU GOING TO DO WITH
THE GRASSLAND BENEATH THE ORCHARD?

Believe it or not, this is a key issue in larger or field orchards. As much as you might think an orchard is about trees and fruit picking, the biggest management issue in larger orchards can be the grass or vegetation that grows beneath the trees. Do you cut it in some form, be that with a mower, scythe or strimmer? Or do you graze it with animals? Either option has considerable implications. If you are going to mow it you don't need to protect the trees except from rabbits, and possible deer, but you have to invest capital in a mower that requires fuel, storage, servicing and maintaining.

If you engage the services of grazing animals, then no mower is required, but as most livestock will eat the trees in preference to the grass, sturdy and appropriate protection in the form of tree guards will be necessary. As we saw in Chapter 9, such guards can be time, resource and capital intensive and can, at the initial outset, cost three or four times what the actual tree costs. The site may well dictate your decisions: livestock may be totally out of the question; there may be no deer or rabbit threat at all. Your wider plan will dictate if domestic livestock are even part of your design.

Some long grass is great for wildlife and definitely to be encouraged. In my own orchard I have skylarks nesting in the grass and, in winter, snipe are often seen skulking around in the uncut areas. But left for too long, and an orchard will start to scrub up and your beloved plot will turn into a native woodland, your fruit trees overrun by thorns, brambles, birch, hazel, oak and those other pioneer species that would return much of our landscape to woodland if we humans were not to continually frustrate their desire to re-wild the countryside.

Managing the sward

This issue of managing the grass is key. My field orchard, at around an acre, grows a lot of grass and for the first three years it was a real headache and a lot of time was spent scything and mowing, and whilst the immediate area around the young trees was kept free of vegetation for weed control, it did at the height of summer resemble saplings in a sea of waist-high grass: just grass, grass and more grass.

An orchard will quickly turn to scrub and woodland, here a young oak has germinated in the mulch around a young apple tree

An orchard ought to be at least a dual land use situation, hence the idea that a yield of some sort ought to be possible from the grass. As we saw in Chapter 5, there are options for yields in between your trees so using the grass growth for something is also common sense. Grazing is the traditional practice in orchards, these days usually sheep, although in some places like Herefordshire, famed for its cattle, beef and orcharding did go hand-in-hand with the beautiful Hereford cattle grazing beneath the standard fruit trees. The problem is that sheep and, even more so, cattle, like eating

In the early years grass was a huge management issue in my young field orchard

Nancy and Woody, my eventual solution to managing the grass

fruit trees more than they like eating grass, so unless the trees are well protected the livestock will eat your orchard.

Tree guards are the obvious solution and a wide range of designs and materials are out there (as we saw in Chapter 9). Personally, I ideally prefer my orchard to look like an orchard rather than a series of trees imprisoned in cages. However, weighing up the short-term cost and labour of tree protection, set against the saving for years to come of not having to worry about managing the grass, should be considered. I have 120 trees in my field orchard so individual protection seemed, at the outset, unrealistic.

Having grown up myself in a slightly Dr Dolittle type household with all manner of animals, including bottle-fed pet lambs, I thought my own children ought to have similar opportunity, so in very early spring 2019 I got them two pet lambs to bottle rear. End result: two very tame, amenable and relatively easy to handle sheep – pet sheep as potential lawnmowers! By late 2019, the initial plantings were approaching six years old so the larger trees only required a wrap around the trunk as protection and I did concede to building 40 or so weld mesh guards around smaller, younger trees, convincing myself that years of not having to worry about the grass was worth the short-term effort. So now the sheep, Nancy and Woody, do an excellent job of keeping the grass in check. Stocking rate is obviously crucial; I only have the two sheep in an acre so making for a very low intensity, extensive form of grazing. Even at this level it is important to have at least another area that they can be moved to, allowing rest periods and to minimise the likely problem of internal parasites; moving to fresh ground breaks their lifecycle. Keeping livestock is a major commitment but means regular and effective sward management and scrub control can be achieved without the need for expensive machinery.

Being the orchard grassland manager is a job for all seasons

The one animal I have not had a chance to try out as orchard grazing tools is geese. In some places they were historically used in orchards and, in theory, could do a good job, as unlike sheep and cattle they won't strip the bark or tear off branches. As I don't live at my field orchard, it is simply not feasible to keep geese which need locking up every night and letting out every morning to prevent the unwanted attention of foxes.

Livestock aside, managing grass constitutes work of one sort or another: mowers, fuel and time. But one approach that has made something of a comeback in recent years is the tradition of the scythe. Yes of course it is work, but is also exercise, a way to burn off a few calories, and mastering an ancient skill is extremely satisfying. The old English scythe was a cumbersome tool whereas the modern, popular, ultra light-weight Austrian scythe is a far superior tool altogether. I find an hour's scything early in the morning a great way to manage areas of longer rank grass that the sheep find unpalatable. When I first learnt scything a decade or so ago, the greatest revelation to me was that you do not wait until the sun is well up and the grass has dried, as we might do when mowing the lawn. The early morning, with dew on the grass, is the ideal time to scythe; the wet acts as a lubricant making the blade slice through the grass far easier than it would cut through dry grass. So I find the best approach is to start early, do an hour or two and then pack up by mid-morning. I cannot think of a more eco-friendly form of grass cutting, using a precision cutting-edge hand tool and getting fit into the bargain – the best of all worlds.

*"The Fruit is what
really matters, not how
gnarly or beautiful
the apple tree is."*

Aiden Wilson Tozer

CHAPTER 12

Enjoy the Harvest

Returning to the beginning when you were planning your orchard, one of the primary drivers may have been what form of production you were looking to achieve. Whether that be a selection of mixed fruit for the household's usage or a specific line of production perhaps related to a hobby or business venture, be it juice, cider, fruit vinegars or other preserves.

The harvest can be viewed as falling into a range of categories: that which can be stored as it is, that which requires some form of processing, and surplus that can be sold or shared.

Understanding the nature of the crop and the work that it will entail is key to a successful harvest of any sort. Do you need certain equipment and is there a degree of processing involved? Does your crop need to be dealt with straight off the tree or can the fruit be stored for later use? Examples include certain plums like the 'Yellow Pershore' that need processing within 36 hours of picking, as they really do go off that quickly, whilst perry pears are considered best left to blett for a couple of weeks or so prior to juicing. So, understanding timings will help you manage production and processing.

HOW TO PICK

Far too much fruit is picked when it is not yet ripe and equally, way too much is wasted by going overripe and rotting. So judgement and knowledge of how and when to pick is essential. An apple on the tree is ripe when you can place your hand around it and lift it to the horizontal and it comes away from the branch cleanly. If you have to twist or yank it, then it is not ready. To judge if a picked apple is ripe, cut through it and the pips should be the colour of dark chocolate; white or very pale brown pips indicate an underripe fruit.

Much commercial fruit is picked underripe due to the time it spends in transportation and storage; it may be many weeks or even months (in the case of apples and pears) until it arrives on the supermarket shelves. But as small-holders and home growers we should attempt to pick and use fruit aligning with its natural ripeness as closely as possible. Plums, for example, are brought to perfection by warmth and sunshine and need to be picked at just the right time, which is why your homegrown ones will be in a different league to those in a supermarket punnet – picked early to allow for transport and distribution time. The general guide that works best is if the fruit comes away from the tree or shrub easily it is likely ripe; if you have to force it, it probably isn't ready.

HOW TO STORE

Today the notion of laying down supplies and storing the fruit harvest for the months ahead can seem almost bizarre. The global supply chains and year-round availability in supermarkets have meant we have not needed to consider such things for a generation. But before refrigerated storage and globalisation, fruit was a very seasonal crop from the first harvest of earlies in the case of apples, late July or early August, through to the point in late winter or early spring when the stored fruit ran out. So, knowing your varieties as I have said is key, especially those that are great for storing, that can be laid down in trays in a cool, dark, rodent and frost-free place, so that you can eat your own apples and pears well into the winter and even early spring.

There are some very nice wooden apple racks on the market for this purpose. I, however, use the plastic mushroom boxes that are thrown out by greengrocers. With a sheet of newspaper in the bottom, between a dozen to 15 apples or pears can be laid out making sure they do not touch each other. Wrapping individual fruit in pieces of paper is often suggested, however, I refrain because they

Keeping varieties can be stored for use over winter in a cool, dark, rodent-free place

cannot be viewed for condition assessments and you have no way of telling if any have rotted or deteriorated. Using my system, a quick glance every 10 days or a fortnight and any 'bad 'uns' can be quickly removed. When selecting fruit to store, it is essential to pick only the best, most perfect specimens; any bruise, scratch or blemish will quickly lead to deterioration and will not keep. Do spend time selecting the best, and checking they are free of imperfections, and remember only keeping varieties will keep; those earlies are for enjoying off the tree – they will not keep however well you try; it is simply not in their DNA.

Today we can bypass some of the knowledge and the risk by freezing, making the homegrown fruit crop available all year round. Plums are best halved and de-stoned, apples and pears can be frozen sliced, halved or stewed down first, whilst smaller fruit like damsons, cherries and sloes can be popped in a bag and frozen whole. Freezers are a great asset in allowing us to quickly stash abundant harvests away to be processed or used at a later date.

PROCESSING

Juicing

In my own orchard, we juice the majority of the apples after selecting the best eaters and cookers for winter storage. Having access to a community owned fruit mill and press, and owning a small pasteuriser, enables us to produce 80-120 bottles of homegrown apple juice every autumn. In the 10 years since we started pressing our own, we reduced our purchasing of orange juice massively, which of course makes a considerable difference in terms of carbon footprints. Apples grown a few feet away or oranges from California or the Mediterranean? Having a stock of home-bottled apple juice is also a great asset as we regularly send visitors, friends and family home with garden produce and a couple of bottles of home-pressed apple juice.

Juicing fruit is a two-part process; one cannot simply squash an apple to get the juice out of it. So the first part of the process involves breaking up the fruit into a mash or pulp; this begins to separate the juice from the fibres. It is this mash that is then pressed under pressure to liberate the juice. At the most basic level, fruit can be pulverised by means of bashing it with a large piece of wood, but the best system is a fruit mill, sometimes known as a 'scratter', a purpose built machine that rather resembles a food-grade garden shredder. Fruit is poured into a hopper (a large, pyramidal or cone shaped container) at the top and almost immediately comes out the other end in an almost puréed form. This fruit mush is then put into a press and wound down under great pressure to squash out the juice.

At the end of the process the pomice or leftover pulp should be so dry you can squeeze a fistful of it really hard and still maintain a dry hand, proving you

1 The first part of the process involves breaking up the apples, here using an electric mill

2 The fruit pulp is then pressed to squeeze out the juice

3 The end result: one hundred percent pure apple juice

4 A home pasteurising set-up enables juice to be stored in bottles for a year or more

have extracted the maximum amount of juice from your fruit. Once you have your bucket or barrel of juice, you can either drink it fresh, freeze it, pasteurise it or ferment it.

It is worth stating that whilst home-pressed fruit juice is a wonderful wholesome product, juicing does involve leaving behind and usually throwing away the fruit fibres; so juicing is great, but from a health point of view, eating the fruit whole is better still. Juicing is ideal for processing surplus that may not otherwise keep, and turning it into a form that will keep, for if pasteurised, a shelf-life of two years plus is possible. I have a purpose made pasteuriser that couldn't be easier; bottles are filled, allowing for an air gap, and then lids are screwed down tight. They then sit, 14 at a time, in the water bath, which is somewhere between a slow cooker and a village hall tea urn. It has an adjustable thermostat and a timer, so set to 70°C (158°F) and left for 20 minutes and the bottles are pasteurised and will keep for up to two years, although we always use it up in the following 10 months or so, ready for the next year's crop.

Fermenting

Cider and perry were once a huge and integral part of rural life. In some regions, the majority of the apples and pears were used for its production. Most farms made their own and counties such as Devon and Somerset along with the three counties of Herefordshire, Worcestershire and Gloucestershire abound with tales of rural cider folk, cider making and drinking. The thing about cider is that it is not to everyone's liking, and within cider there are some superb brews, some absolutely atrocious rough stuff and everything in between. Thus to claim one likes cider is a sticky wicket to play on. By all means, have a go at making your own. Whilst making your own is a great use of surplus apples and good fun, the results will vary year on year; I have never progressed beyond the scale of a few demijohns a year. Factors like sunshine levels, the exact mix of apples you use, and the weather in general, all alter the character and sugar levels in apples so your cider will be bespoke.

The subject of making cider, or for that matter perry, is one of those rural subjects where every countryman

From apple to cider with nowt added at all, homemade and eco-friendly. In its purest form cider is just fermented apple juice, using the natural yeasts in the skin.

or woman seems to have their own, and usually different, opinion. "You don't want to be doing it like that" is an oft-heard comment among cider folk. At its most basic, cider is simply apple juice allowed to ferment. That is it. Nothing added; natural yeasts in the skins cause the fermentation process. However, many makers do also add a special cider yeast which is considered to make for a more reliable, uniform product, taking the more 'hit and miss' elements of natural yeast usage out of it. I have tried both methods over the years and have to agree that adding in a bought yeast does seem to make it more reliable. But equally, there are some very good cider makers who advocate just juice and good old wooden barrels. So it is a subjective judgement as to what constitutes 'proper cider' and just like with the apples themselves, we have our individual palates and preferences.

Another great aspect of cider making is its green credentials; being a cold ferment it uses very little energy in its production compared to beer, which needs heating up. A small-scale home cider-making operation using local apples can have a minuscule carbon footprint.

Vinegars

Apple vinegar is essentially oxidised cider, i.e. cider that has received too much exposure to oxygen and as a result, gone acidic. As a product, it is totally natural and if marketed well, can be the most profitable form of juice. I have made my own by simply filling a demijohn half full of apple juice, sticking a wad of cotton wool in the neck, and leaving for six weeks. We have pickled our own onions and beetroot in our own apple vinegar and used it to make our own chutneys and salad dressings. An easy to produce, homemade product; go and look in a health food shop and see what they are charging for a small bottle of cider vinegar.

Other preserving methods

Historically there was a range of ways fruit was preserved to make it last beyond its natural season.

Apple was much used in the jam industry as a bulking agent, being added to other more valuable or expensive fruits; it also contains pectin and helps with the setting process of jam. By adding a proportion of apple, a far cheaper fruit, it made the main ingredients go further, whilst the blandness of the apple did not significantly impinge on the intended flavour of the main fruit. Apple jellies, often made from those bitter inedible crab apples, have long been considered a fine accompaniment to cheese and meats.

An unusual and far less well-known preserve is black butter or black apple butter as it is sometimes known, also called apple stroop in the Netherlands. Somewhat like a jam, it consists of stewed down apples, often cider varieties, with cider, sugar and spices like cloves and cinnamon. It makes a spread that

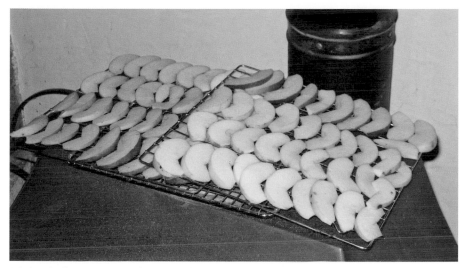

Dried apple slices are easy to make on a woodburning stove and once dry will store for months

can be eaten on toast, biscuits or even on its own. Recipes vary and it is not a butter nor does it contain any but is so called due to its smooth buttery texture. Thought to have originated in the low countries of the Netherlands and Belgium, it is associated with various regions here such as the West Midlands.

Other culinary uses saw fruit used in a wide range of baking; there were various versions of apple cakes, plum tarts, Warden (pear) pies and fruit crumbles – their demise is a reflection on the way hearty puddings have fallen from fashion.

Another form of preservation is drying; apples especially can be finely sliced into rings and dehydrated to make a healthy snack or 'apple crisps' as some like to call them. There is even a variety, the 'Herefordshire Beefing', that was claimed to be a specialist drying apple, presumably due to a low moisture content and dry flesh making it easy to dehydrate. Today state of the art dehydrators can be bought enabling all manner of fruit and vegetables to be dried for preservation, a great way of enjoying the orchard harvest all year round.

SELLING

In horticulture, there is a phrase known as 'The Grower's Dilemma' where many of our farmers and growers, who put in all the work, make less profit than the middle men and women. Hence the only way to make a decent return has to be to cut out the middle person and sell direct to the customer, keeping one hundred percent of your takings. This is still easier said than done. Fruit tends to be a bulky, in some cases heavy, crop with a relatively low return – hence knowing your market, and harvesting, packing and transporting it all needs careful consideration.

Sometimes a garden gate stall is as good as anywhere to sell surplus fruit

My own orchard does not make sense commercially, as all the different varieties with their staggered ripening and harvesting times make it utterly inefficient. That is not to say I haven't considered a range of monetary ideas, even if only to make a small return that could be reinvested in the orchard; 'The Unusual Apple Company' was one such idea. As I have said, people are more and more interested in local produce and where their food comes from. People also like a story, so the idea of unusual apples that one would never see in the shops and the interesting stories associated with them might appeal to a certain, if small, niche market. I live close to the town of Ludlow, renowned as a foodie centre of national repute, the sort of place such an idea might just work. However my rough calculations and investigations suggest that by the time I have picked my crop, packed it, driven it and tried to sell it direct at a market stall or similar, it simply doesn't appear to stack up. Overheads and a day of my time would yield little in return. A stall at the orchard gate and an honesty pot might well yield a similar net percentage return.

So I have not cracked selling my produce, but then my orchard was never set up with any commercial intentions. It is a collection, a hobby, and any business venture, if it ever does happen, will be an aside. Some of you may have far better ideas and success at making an orchard yield in a monetary sense than me, and I wish you every success. One of the reasons so many of our old orchards disappeared was that they simply didn't pay any more. Proving some sort of financial return makes for a compelling case for the planting and preservation of orchards, which perhaps takes us back to cider and perry, for with the growth of interest in artisan drinks, cider makers are the most obvious outlet for surplus fruit.

Sharing

Business ideas and talk of money aside, there is something nice about sharing a bounty by simply giving away a lovely bag of homegrown, unsprayed fruit to friends and family, or even strangers. It is simply a very human thing to do. 'Give and you shall receive' is after all a lovely idea. I have often given fruit away and later had other plants and produce in return.

In my part of the country, the sight of boxes of apples at the end of drives or garden gates with 'HELP YOURSELF' signs is fairly common in September and October in a bountiful year. In the town of Tenbury Wells, the community orchard have their own Apple Shop, the difference being this one doesn't sell tech gadgets, indeed it doesn't sell anything, as the fruit is free – a stall just off the high street where you are invited to take a bag of apples for free is a lovely embodiment of the best of human nature and sharing the crop with others.

But it is not just our fellow men, women and children that it is nice to share the bounty with. In the early days of our orcharding I felt very guilty about not utilising every last fruit, but over time I have realised that allowing a degree of the surplus to lie on the ground as windfalls is a great contribution to our local wildlife. Now, when that point in the winter comes when the temperatures plummet, and the flocks of fieldfares and redwings descend on the orchard and garden and clear away the fallen fruit, it is an event to be cherished and looked forward to. The fruitful season is enjoyed by a vast range of our native wildlife.

Foxes and badgers change their diets to include a considerable amount of fruit in late summer and autumn. Butterflies, such as the red admiral and comma, love the fermenting juices of windfalls. Small mammals like rabbits and mice, as well as a wide range of birds, will all come to feast on fallen fruit. So my guilt at waste or fallen fruit has subsided in later years, as like so much in nature, nothing is truly wasted.

I also like to continue an old tradition that you never completely clear a tree and, when picking, always leave at least the last two or three fruits on the tree for whatever or whomever may come along after me.

The Tenbury Wells Apple Store – no money required, photo by Charlie Wilcock

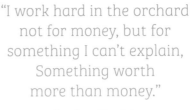

"I work hard in the orchard
not for money, but for
something I can't explain,
Something worth
more than money."

Stephen Herrick

CHAPTER 13

Managing for the Long Term

Whatever level of support and protection you opted for, it is important to regularly check that guards, stakes and ties are doing their job and not hindering the tree by chafing on it, or growing into it. So regularly take a walk around your trees with the specific purpose of doing that bit of tending. Loosening ties, maintaining guards and generally checking up on your trees is an investment of incalculable worth.

APPLIED OBSERVATION

Never underestimate the value of just wandering through your garden or orchard, perhaps with a cup of tea in hand, just looking at it all; you will learn so much by doing this regularly in each season. Just seeing what does well where and what does not; which plants are surviving, even thriving, and which are not, you will gain a mental picture of the dynamic of your plot, from which lessons can be drawn and tweaks and alterations made to the system.

I spend a considerable amount of time just wandering around, taking it all in. To others it may look as though I am not being very productive, whereas to me, the exercise is invaluable. A garden or orchard is forever evolving, and there are always endless amounts more to learn. I keep records of my fruit trees, of how well they perform each year, and know the layout of all my trees and varieties almost off by heart. So, regularly wander through the orchard, up and down the rows of trees; observation will highlight trees that are flourishing and those that are not, adding to that ever deeper tapestry of the knowledge of your orchard. But don't just build up a fair weather picture, don a good hat and coat, and go walk about in the rain or even snow; you will make as many and as equally useful observations from getting a bit wet or cold as those you may make on a sunny day. Note how well your soil soaks up water, where it lies; do you have a problem with compaction or poor drainage? Fruit trees tend not to like standing in saturated ground for long periods. So go out regularly in all seasons and conditions to build up that encyclopaedic knowledge of your site. Aside from keeping notes and paper records, technology enables us to film and photograph the orchard; even a moderate smartphone will take high quality images. So record your trees and their crops through the seasons and from year to year.

Although I do not live at my orchard, the daily requirement to check the sheep and our cattle on the rest of the land means twice daily visits to the field, so I walk the orchard most days and, in doing so, build and edit that constant mental picture of how it is all going.

In the autumn of 2020 following a wet spell, the number of fungi that came up in the orchard seemed way more than I had ever noticed before, causing me to wonder if the land was recovering, accepting my limited use of herbicide in those first two years. In the five years since, the land has been chemical free and perhaps nature is beginning to respond to that recovery. Whilst I am not sure what several of the fungi were, it seemed like a good sign; seeing them was again a product of that wandering around.

Never underestimate the value of just wandering and looking

Truly getting to know your plot means observing it in all weathers

PRUNING

There are countless books on pruning and, in truth, it is a very hard subject to learn from a two-dimensional page when a tree is such a three-dimensional structure, so the best advice is go on a course or go and learn hands-on from someone doing it. However, we should perhaps discuss the reasons why we prune and what we hope to achieve by doing so.

There are four main reasons why we prune fruit trees:

To get better quality fruit – not necessarily quantity but quality. It is far better to get 100 fully ripe, fully sized apples, than 200 unripe specimens the size of a golf ball. This introduces the concept of thinning and selecting for quality fruit.

The second reason is to let light and air into the tree. It is sunshine, i.e. light, that ripens the fruit, and good air flow blows away spores and disease. A healthy fruit tree should have a light, open, airy framework and not be a crowded, dense structure with poor air flow and lack of light in the middle of the tree.

Thirdly, we might prune a fruit tree to keep it a certain size so it does not romp away and dominate our garden or plot. We may prune it so it doesn't impinge on the path, shed, greenhouse or washing line. Thus, size-restrictive pruning might be purely practical in managing the space available.

Then lastly we prune to remove the three Ds: dead, diseased and damaged. So in a tree we might look to cut out snapped and broken branches, any dead wood, and certainly any diseased wood. Common fruit tree diseases like canker have traditionally been managed by cutting out any sign of it. The subject of deadwood is a contentious one and, as we shall see later, is highly valued for wildlife, so that the traditional practice of chopping it all out is something we may well not endorse so freely today.

In understanding fruit trees we have already seen that they are un-natural creations, being two entities spliced together, but they are also, once established, a managed entity. We manipulate them for our own ends to produce as much, and the best possible, fruit for our consumption. Pruning trees is about balance; being deciduous, a fruit tree, when left to its own devices, will have a limited amount of energy derived from photosynthesis and soil nutrients that are available to it. From the point it breaks bud in spring, to the point it drops its leaves in autumn, if left alone it will divide

The classic wine glass or goblet shape is the most common form achieved for garden and amateur orchards

that energy between fruiting and growth, i.e. growing more tree and pro-ducing a fruit crop. In pruning we can alter that balance in the tree and get it to use that energy for more fruiting or more growing. So we are dealing with an artificial, managed situation. Growing fruit trees in an orchard is not like wild trees growing in a forest.

There are a range of shapes and forms of fruit tree as we saw in Chapter 9, but for field-grown trees the most popular and most often referred to is the 'goblet' shape. Imagine the trunk as the stem on a wine glass or goblet, and then the crown of the tree resembles the glass or cup shape: an even rim of branches heading away from the centre of the tree with an open airy centre to it. A goblet pruned tree will usually have that stem and then between three and seven main branches, off which all the smaller branches and fruiting growth comes. Go and look at any old garden apple tree and I can almost guarantee this rings true; there will be a trunk and then it will split into three, four, five or more large branches.

It is also a botanical fact that horizontal branches tend to produce more fruit than vertical ones, which have a tendency to produce more growth, hence favouring the horizontal over the vertical is one of my mantras when pruning fruit trees. In general terms, this usually helps keep trees a manageable size. The first job I nearly always do when pruning apple trees is cut out those vertical

In recent years there has been a move toward summer pruning, especially of trained and bush forms of tree

shoots going straight up for the sky and then I look at the more horizontal branches and select the best to give the size, shape and form of tree I am after.

A common pruning rule, particularly with a neglected tree, is that no more than thirty percent of the tree should be pruned out in any one year. This is good advice, as any pruning is a shock to the tree and, in the case of very old ancient trees, they could be shocked to death. Equally, over-pruning a tree of any age can result in an explosion of water shoots, those long wispy regrowths that make a tree resemble a giant lavatory brush. Such growth will not fruit for a year or two at least, and leads to that dense over-crowded structure that we want to avoid. Little and often is by far the best pruning approach.

Any books or instructions on pruning pre-1990s will always make reference to pruning being a winter activity, yet there is a more modern school of thought that advocates summer pruning. The tradition of winter pruning perhaps had more to do with the farming calendar than what was actually ideal for the trees. In the days of large rural workforces, once the autumn harvest was all in and safely stored away there was potentially a surplus of labour at what could have been a quiet time of year. So it fitted the agricultural year to send them out to prune the orchards. In the case of apples and pears, you can actually prune them on any day of the year so it was by no means essential to do them in that post-harvest window.

Stone fruit, those members of the *Prunus* family, plums, damsons, gages, cherries and apricots, are a different matter. These have to be pruned between May and September when the sap is up and they are in leaf. This is because of a fungal disease called Silver Leaf which, like most fungi, produces fruiting bodies and spores in the autumn; exposing open wounds from pruning to the spores in early winter is to invite infection. Silver Leaf, once in a tree, is fatal and although it may take a few years to fully kill a tree, it cannot be treated. The technique of pruning between May and September is two-fold. Firstly there aren't many silver leaf spores around at that time of year and secondly, if there are a few spores, fresh cuts while the sap is up will bleed, in theory flushing off any spores that may land on cut surfaces.

By cutting back this year's extension growth, light is let onto the fruit in late summer and overall vigour reduced

Apples and pears do not suffer from the same diseases as stone fruit and are, as a general rule, more hardy and resilient. Pruning any tree when in full leaf is harder on the eye, as the structure is obscured by leaf cover and seeing what you are trying to achieve can be more difficult. However, it is generally acknowledged that winter pruning has a tendency to produce more vigorous growth, whereas summer pruning stimulates more fruiting the following year. In very simplistic terms, pruning a tree in summer when the sap is flowing and it is in full leaf is a considerable shock to it. It is this shock in the tree, thinking it is under attack, that stimulates fruiting, for when under attack a living organism's evolutionary reaction is to want to reproduce before it dies, hence the stress causes the chemical reactions in the tree to produce greater fruiting the following year. This is a slightly simplistic and anthropomorphic view but captures the gist of it. The other advantage of pruning in summer is that the wounds it produces will start to heal almost immediately, as opposed to those made in early winter that will wait until spring to start healing over. Hence, it is my supposition that cuts heal quicker and better from summer pruning.

Certainly, all the trained forms in my garden orchard, the stepovers, cordons and espaliers, as well as the dwarf trees, are all now completely summer pruned.

There are two types of buds on a fruit tree, and it is important to be able to tell them apart. Most fruit trees only produce fruit on second year and older

The author with a couple of productive cordons

The fruit buds are distinguished by their sticky out pointy nature often referred to as spurs

growth, so the current season growth and last year's have growth buds, those that will produce more twig-like growth and leaves, whereas the older wood has fruit buds or spurs. These fatter, more swollen buds that often stick out in a more pointy form are what will open out as blossom, and assuming they get pollinated, produce the tree's fruit. To return to this notion of balance: if we want a young tree to put on growth and get larger as quickly as possible, we don't want it using up lots of energy on fruit production while it is still small, so we can thin out fruit buds, or once any fruit has set, pick it all off when still the size of a pea, thus concentrating energy on new branch and leaf growth. Conversely, when a tree is as large as we want it to be, we want it to maximise its energy into fruit production and not just keep producing more and larger branches. So we look to prune out mostly growth buds at the expense of fruit buds, so the energy goes to produce a good harvest. This latter approach will consist largely of cutting back the current and previous year's growth. In summer pruning, to give a tree a damn good haircut of the current season's growth will not constitute wasting a single fruit; fruit is not produced on the new growth.

Restorative pruning

Those who own old, perhaps neglected fruit trees are often of the view that they must need pruning, but before wielding the saw and secateurs, it is worth taking stock of the situation. Pruning big, old trees is a considerable human endeavour in time, labour and, potentially, cost. So going back to our four main reasons why we prune: if we are not looking to obtain a commercial crop and the tree isn't causing us issues of space, we should question whether it is worth the effort. Many old trees will sit there for decades quietly, as a hugely valuable habitat for a host of wildlife, blossoming, producing enough fruit for you and the local wildlife. As such, unless there is good reason to do so, I take a very minimal intervention approach to these old and ancient fruit trees.

For restorative pruning, first make several saw cuts to open up the frame of the tree

A well-made saw cut on young-ish trees will in time heal over completely

Never underestimate the resilience of fruit trees; this ancient specimen is still healthy and cropping

KEY POINTS TO REMEMBER WHEN PRUNING

- Prune out diseased wood wherever possible.

- Prune out branches that cross or rub on each other, as this causes damage and acts as an entry point for disease.

- Favour the branches that are going to give you the shape of tree you want; where two branches are close together and compete for the same space and light, remove the weaker one.

- Thin out over-congestion; remember we are looking for a light, airy tree, not a crowded dark one.

- In the ideal fruit tree, no two branches or shoots should touch each other and there should be good air flow to all parts of the tree.

The biggest case of mortality in these veterans is the sheer size and weight they have grown to, thus on wild and stormy nights they often topple or break up, hence we introduce the notion of rebalancing. If a very large old fruit tree appears rather one sided or has a significant lean, we might lop off the offending limb or part of it, in order to lessen the likelihood of its own weight pulling it over. But beyond that, I am inclined to leave them to grow old gracefully in the landscape, providing a fantastic habitat, and to concentrate my energies on establishing and developing younger trees.

MULCHING / MUCKING

Whatever your soil was like to start with, you can enhance it on a continuing basis by adding organic matter in the form of mulch. A thick annual application will cap in the winter moisture in the soil, suppress unwanted weed growth and slowly feed the tree as the mulch breaks down and is incorporated into the soil by worms and microorganisms. The nature of mulch will be dependent upon what you have access to: farmyard manure, woodchip, grass clippings, soiled hay or straw; all are usable as tree mulches.

There are pros and cons to each; farmyard manure needs to be well rotted as it can be too acidic when raw, and if not rotted, it can be a huge

A good application of mulch in late winter / early spring is highly beneficial to the tree

source of grass and weed seeds. Soiled or waste hay also tends to be full of grass seed. My favourite is woodchip as it stimulates that mycorrhizal activity, breaks down slowly and is largely free of weed seeds. This feeding of the soil, boosting worm activity and that mycorrhizal life, goes a huge way to enhancing the health of your soil into the future, however poor it may have seemed at the outset. Late winter is my preferred time to do this particular job, capping in the winter rain and getting a good smothering layer down before spring growth commences.

CHAPTER 14

Working with and for Wildlife

Whilst an orchard might be primarily for fruit production, it can also be a habitat, a home to local wildlife. With careful planning and the right management, you can enhance it for the benefit of the flora and fauna and they, in turn, can be orchard allies in pest control and pollination. For too long the word pest has been used in association with wild creatures. Whilst a few like rabbits and deer can certainly be a real problem, far more wildlife can be beneficial in the orchard, controlling problem insects, clearing away waste or fallen fruit. And then there are the benefits that we as humans obtain from seeing and being in amongst wildlife.

BUILD IT AND THEY WILL COME

Nature is wonderfully resilient and it never ceases to amaze me just how quickly it can come back when we allow it to do so. When I purchased my field it was a big expanse of badly overgrazed grass with virtually no habitat value at all; the hedges were planted along with the young fruit trees and the grass was allowed to grow. Quickly we were finding frogs and toads in the long grass, despite no nearby ponds, spiders, beetles and other invertebrates seemed abundant, and a kestrel regularly hunted overhead. An abundance of birds, from yellowhammers and wrens to robins, tits, magpies, pigeons, skylarks and even snipe, were seen in the orchard. The commonly used adage that 'if you build it, they will come' is absolutely true when it comes to habitat creation.

Deadwood has always had a rather bad press, being seen as dead, and perhaps diseased, and so something to be disposed of, whereas it is, in fact, a vital part of the lifecycle for so many creatures. Deadwood in the orchard is a good thing, so although we cannot recreate an ancient hollow or decaying trunk, we can create artificial versions. In my orchard, I knocked in a stake and then wired some branches from a local tree surgery job in an upright position so, as they decay over time, they will provide invertebrate habitat that would otherwise be absent from a new site. Many overlooked insects like micro-flies need dead and decaying timber to breed in and, in turn, perform valuable pollination, so aiding their existence in your orchard is a benefit for you and them.

Despite no nearby ponds, in no time at all both frogs and toads were to be found in the long grass

Another feature lacking in a new young orchard is natural nest sites: those cavities in hollow trees in which birds can build their nests. This is easily rectified by putting up bird boxes, but of course young trees are not of a size or physical stature to accommodate having a box hung on them, so posts knocked in around the orchard, with boxes mounted on them, provide that opportunity. Aside from doing something positive for the birds, there is a great pest control element as blue tits and great tits eat vast amounts of caterpillars, something like 1000 a day when feeding chicks, so having them nest and rear a family in the midst of your fruit trees is free, organic pest control that absolutely works. I have stood and watched blue tits nesting in a box, three feet off the ground, going backwards and forwards, a few yards between the trees and the box, beak full of caterpillars. A true win-win.

An angle shades moth was just one exotic summer find

Wildlife campaigns tend to focus on large attractive animals and the little creatures are often overlooked, but so much of the work in ecosystems is done

Bird boxes mounted on posts throughout the orchard is a very easy project with multiple benefits

by these largely unseen heroes. So, giving them a helping hand is vital and easy to do. Piles of logs and sticks, compost heaps, and mounds of decaying grass are all home to the little critters. It has been recognised elsewhere but the loss of insects has been catastrophic. A very everyday observation: as a child of the 1970s, in the back of the family car on a country drive, the windscreen would have been carnage with dead insects, yet to do a similar drive today the windscreen is almost clear. A whole mass of insect biodiversity has disappeared, with us barely noticing.

WILDFLOWER MARGINS

Therefore, anything and everything we can do to help the micro-wildlife is extremely valuable, from habitat piles to sowing pollinator strips or wildflower margins, all of which give homes and feeding stations to those that remain, and in return, they boost pollination, so aiding our fruit production. Even while suffering these catastrophic declines, insects do breed quickly and in potentially huge numbers, so they have the potential to bounce back quickly once the habitat is right. The main reason for their decline is habitat loss, as vast agricultural fields, heavily sprayed with chemicals, are utterly inhospitable to our insects.

Really simple additions can make a big impact; my children and I have built a range of insect houses and bug hotels and to see them used within their first year, teeming with various types of solitary bee is amazing, and an example of how giving nature that little helping hand makes a big difference.

Sowing pollinator strips or wildflower margins is hugely valuable on any scale; they will draw in a wide range of insects that in turn will carry out those pollination and pest control duties. So if it is a few pot or plug-grown wildflowers

Tawny mason bee and dagger fly are just two of the unsung heroes of pollination, photos by kind permission of Steve Hughes

or cultivating and tilling a strip into which a wildflower mix is sown, adding diversity to your orchard will reap many rewards. In addition, there is a whole subject around companion planting and the potential benefits of polycultures and how the presence of one plant can be beneficial to others.

As discussed in the earlier chapter on sward management, it is necessary to manage grass growth in order to prevent your orchard from ultimately scrubbing up, but a rotational approach to having different heights of sward will give a home to different wildlife.

Traditional orchards are now considered an extremely high value habitat and with large, old trees, nooks, crannies and crevices, deadwood and established hedgerows and boundaries, a wealth of wildlife can find a home and food to eat. In a newly established young orchard, most of those features do not exist, so the habitat value is far less. However, we can take a range of measures to recreate some of these features and draw more wildlife into our new orchard.

BEES

Honeybees are the poster girls of pollination and yes most of them are girls, but in truth a vast amount of pollination is done by less glamorous, less well-known insects and it is their almost unnoticed declines that should give us cause for concern. I am no entomologist and work on the 'every little helps' principle that if you create good habitat, creatures will move in and live, breed, survive and hopefully thrive there, even if I don't know their names, species and orders. It is an interesting exercise in spring to take 10-15 minutes to just stand alongside a fruit tree in full blossom and watch what visits the blooms. You may be amazed at the range of tiny flies, wasps, solitary bees and beetles carrying out pollination, most of which you probably never even knew existed. This mass of biodiversity that is almost completely unstudied to a point where very little is known about the scale of decline, does in truth perform a hugely valuable role in pollinating food crops, in our case fruit trees.

Honeybees are very temperature sensitive, and do not come out on cold days, so early blossoming fruit like damsons and some plums are almost entirely pollinated by non-bee species. As I say to my kids, "the honeybees get up late, go to bed early and don't come out at all if it's too chilly," something which certainly seems to be the case here in the Shropshire Hills. Honeybees have got a lot of publicity in relation to their disease problems and declines but we must learn to value all our insects equally and do what we can to help them. Bumblebees interestingly are descended from alpine species and their furry, fuzzy bodies enable them to fly in much cooler temperatures than their honeybee cousins. Hence, early and late in the season, as well on unseasonably cool days, here in the Midlands you may well see bumblebees out and about when there is not a honeybee to be seen.

Natural non-intervention beehive in the orchard 2020

Having dabbled rather unsuccessfully with bees in conventional hives, I now have a couple of non-intervention hives, essentially wooden boxes made of scaffold planks that are imitating hollow tree trunks. The bees moved in by themselves, via a swarm, and seem to overwinter successfully with zero intervention on my part. As none of my family are particularly keen on honey, we are happy just letting them bee! Taking the view that bees make honey for bees and the concept of taking much of it and giving them some cheap white sugar instead, well is it any wonder bees are in trouble? I love having bees in the orchard and am genuinely very fond of mine; I find them fascinating creatures and often spend 10-15 minutes just watching their comings and goings. So yes I have bees, but am a non-interventionist. My orchard is remote, a good half a mile from the village and the nearest house, so they behave as would wild bees, undisturbed by human hands.

LARGER WILDLIFE

To some extent, the big wildlife may be determined by the part of the country in which we live, but for me there is one talismanic creature with which I am almost obsessed and seeing one never ceases to raise the spirits and create a sense of awe. Here in the Shropshire Hills, we are fortunate to have red kites. If ever there was a good news bird story, this is it. Once persecuted to the very brink of extinction they have, with a lot of help from dedicated conservationists, made a remarkable come back and are living proof that if we put our minds to it and there is a will, we can bring back amazing wildlife.

I would not remotely class myself as a birdwatcher, but there is something mesmerising about this bird: smaller and more delicate than its more common cousin the buzzard, there is something about the silhouette and then the way

THE NOBLE CHAFER

The Noble Chafer is one of the UK's rarest beetles and has of late become something of an emblem for the conservation of traditional orchards. These stunning metallic green beetles are around 2cm long and the adults only emerge for a brief period in midsummer where they may be occasionally observed feeding on white flowers. They spend two to three years as larvae living inside old fruit trees where the grubs feed on the decaying timber, emerging as adults simply to breed and start the cycle again. The loss of old orchards and the obsession with tidiness and clearing away deadwood have been key factors in their decline.

they hang in the sky and pivot their distinctive forked tail, as if balancing effortlessly on air.

They are still not plentiful in this part of the Midlands but sightings year on year are seemingly on the up, and to stand in my orchard with a red kite gliding in, through a clear blue sky is about as good as it gets from a wildlife watching experience. Kites have a long history of association with people – as scavengers they have made the most of what we humans often waste. My family and I were sitting in the orchard one sunny summer's day eating sandwiches when a red kite came in to check us out, buzzing at little more than a dozen feet above our heads. Sadly for him or her, it did not yield a sandwich on that occasion.

Whilst there is plenty to observe in the orchard on the ground, in the trees and hedges, there is also, like the kite, wildlife to be seen overhead. The local kestrel being an almost daily sight, making the most of the long grass, is an example of how such a simple act as letting the grass grow long suddenly provided a feeding opportunity that was not there when the field was an over-grazed sheep ranch. In spring 2021, kestrels nested in a box I put in a large ash tree just outside the orchard, so a simple case of putting up a box and creating the right habitat yields a wildlife win. Whilst I have not seen one, I do have a hope-cum-fantasy that after dark the occasional barn owl perhaps frequents my orchard, gliding beneath the trees in search of a vole; I have seen them in the fields and lanes nearby so they are in the vicinity and in theory could come our way. Even when the orchard was just six years old we had goldfinches nesting in the apple trees, their intricate nests of moss and feathers wedged in the forks of the trees. Sadly they were not disguised well enough, for it seems the crows and magpies had also spotted them and we found them torn asunder-raided for the eggs or nestlings. Life in the orchard can be brutal as well as beautiful.

Ecological tidiness disorder is a major cause of wildlife declines. As a species, many of us humans have this urge to tidy up nature and the countryside, whereas much of nature rather likes a bit of a mess – the forgotten corners and shabby overgrown waysides are where wildlife finds refuge. When we needlessly tidy it all away, we take away yet more food and homes from wildlife, large and small. So when it comes to an orchard or garden, resist the temptation to be too tidy; leave overgrown areas, piles of wood, patches of nettles, or a few thistles, as they all have roles to play in making a rich and varied place for all creatures, great and small, to thrive.

A point I come back to again and again, and one that is found repeatedly in nature, is that diversity is king. Monoculture is almost never encountered in natural systems. The general principle of creating a rich and diverse environment in which pests and their natural predators are present ought to create a more natural balance and whilst yes, some pest species will continue to be present, natural checks and balances ought to come into play. One of the lessons of modern agriculture is surely that monoculture crops lead to monoculture pests. Diversity really is nature's way of all things, a lesson we would do well to heed. The great potato famine of the 1840s in Ireland should stand as an all-time reminder against monoculture farming.

"Traditional Orchards are
cultural landmarks and give us
distinctive landscapes. They are
the source of genetic variety, local
recipes and customs. They
are beautiful to be in and
havens for wildlife."

Sue Clifford and Angela King,
The Apple Source Book

CHAPTER 15

Joining the Community

Whilst owning your own orchard is wonderful, sharing one is perhaps even better. One of the lasting legacies of Common Ground creating that renewed interest in orchards back in the 1990s was the subsequent number of community orchards that sprang up: villages, parish councils, residents or just groups of friends getting together to create orchards or in some cases purchase or adopt existing ones for the wider benefit of public enjoyment and bringing people together.

Orchards in the heart of the community embody that permaculture principle that we should grow food right where it is needed, thus saving countless miles and vast quantities of energy in transporting food around. As we have said, fruit is a heavy, bulky crop and so food metres rather than road miles can only be a good thing when local folk can pick fruit in their own communities. Fruit aside, there is huge value in community orchards as shared spaces, a place of greenery, tranquillity, blossom and wildlife, a space to meet, gather, picnic, celebrate and be a community.

One of the joys of such orchards is that no two are the same; some are still young, having been planted from scratch, perhaps with each tree sponsored or donated by local people or businesses. In other situations, groups actually clubbed together to purchase existing orchards and so save them for posterity. The land in some cases is owned by the community, in others rented, leased or gifted, but all share the overriding objective of making orchards accessible to people, bringing communities together and engaging people with blossom, wildlife, trees and fruit.

Community orchards are as, or even more, valuable in urban areas: back to growing fruit right in the heart of towns and cities where the people are. But

Above: The author leading a workshop at a community orchard in Birmingham

Right: Community orchards are a great way to educate the public about fruit, orchards and history

Below: The amount of walls in urban areas adds to the urban heat sink, which means fruit can flourish in towns

whilst space can be an issue in the city, there are also advantages: a profusion of walls that can accommodate trained forms of fruit, the urban heat sink effect reducing the likelihood of frost damage and the possibility to grow a few more exotics like apricot and peach, where those extra three or four degrees between city and countryside may be the difference between fruiting success and failure. The Orchard Project (formerly The London Orchard Project) is doing wonderful work across several major cities engaging people with orchards in the heart of our urban landscapes. Planting, tending, picking, using, educating and nurturing an appreciation and a passion for orchards wherever people live. If this renewed interest in orchards is to have longevity we need people to care about and love orchards.

It can seem a daunting prospect trying to get people together to create an orchard from scratch and there can be an almost obsession with obtaining grant funding. However, one of the best ones I was involved with was based around a piece of church-owned land that the parish was granted a lease on for £1 a year. We then held a cheese and cider evening with a couple of talks on orchards and wildlife. Food and merriment and that one evening alone raised several hundred pounds that paid for all the trees and materials to get the first 50 trees established. All without any need for grants. Tree sponsorship is another great way of funding the planting of new orchards, offering individuals, families and businesses the chance to pay for a tree or two; many people will happily pay say £25 to sponsor a tree in their name. A number of community orchards I have been involved with have used this technique successfully to fund planting of new sites.

Land purchase is obviously the most significant obstacle facing the establishment of a community orchard, but there are all sorts of pieces of land that can be suitable without the need to ever buy them. There is so much wasted, unproductive and often overlooked land around us. Industrial estates, housing developments, wide verges, and just those left-behind spaces between the places where we live and work, could all be made productive with the will and drive of the community.

APPLE DAY

The other great legacy of the work of Common Ground was the establishment of Apple Day. First held on October 21st 1990 with the notion of a special day to celebrate the nation's favourite fruit, 30 years later it has become a firm fixture of the autumn calendar, with apple days, events and festivals throughout many parts of the country. It is an opportunity to immerse ourselves in all aspects of the apple, from the fruit itself to cider, juice, customs, stories and even apple games and apple products. The centrepiece of most Apple Days has become a display of the range and diversity of apples found in the area; the public at large, used to seeing three or four in the supermarket, are amazed at seeing hundreds of different varieties with their amazing names and stories. 'Bloody Ploughman', 'Lady's Finger', 'Ten Commandments', 'Dick's Favourite', 'Catshead', 'Sheep's Nose', 'Hope Cottage Seedling', 'Barnack Beauty' and so on; the names alone suggest dozens of stories of intrigue and quirky tales to be told.

To see several hundred varieties on display is to fully experience the diversity of the *Malus* genus, from large to small, reds, yellows, greens and purples, sharp, sweet, crisp and soft, ones that keep and ones that won't, ciders, crabs, eaters and cookers; the apple is truly a remarkable member of the plant kingdom.

To revive our orchard culture and all that goes with it, from local food, to reducing waste, processing, preserving and enjoying the fruits of our labours, we have to educate the public and Apple Day is a wonderful way to do that. So, October 21st every year is a chance to shine a light on what is so much more than the humble apple.

The vast array of varieties on display at an apple day is a real eye opener to most members of the public

> "I conclude all with a
> catalogue of such excellent fruit
> trees as may direct a gentleman
> to the choice of that which is
> good, and store sufficient for a
> moderate plantation; Species
> and curiosities being otherwise
> boundless and without end."
>
> John Evelyn, 1706

CHAPTER 16

Deeper into the World of Apples

ORCHARD DETECTIVES

There is this long and rich history of people, fruit trees and orchards, none more so than when it comes to all those varieties and the sheer volume and variation of them. As we have seen earlier, some like 'Catshead' and 'Court Pendu Plat' have been around for at least four centuries or more. However, many varieties, including some far more recent ones, have disappeared. It was in the mid to late 1800s that the study and recording of varieties reached a peak, when the learned horticulturalists of the day attempted to capture the wide range of varieties of the time, which they wrote up in Pomonas (often beautifully illustrated books on fruit). The zenith was perhaps an exhibition in 1883, The National Apple Congress held at Chiswick Gardens, when nurserymen and growers from across the country came together and exhibited in the region of 1500 varieties of apple, the most diverse display ever put on in this country.

Those nineteenth century references give us a picture of what existed at the time, at this height of diversity. But a good number, from at least 150 or so years ago, cannot be found today. They do not exist in the national fruit collection nor are they in private collections or museum orchards, nor can they be bought as trees from any nursery. They are effectively the lost varieties.

It may seem odd that a named and known variety can just disappear, yet with a little understanding of their history it is easy to see how this can happen. Of the thousands of varieties once known, many were never widespread, perhaps only existing in a tight geographical area, maybe confined to a handful of villages or a district. Some originated as a local seedling that someone deemed worthy

A plate from *The Herefordshire Pomona* of 1878, arguably the finest work on English apples and pears ever produced. Image by permission of The Marcher Apple Network.

of propagating, given a name and then shared with a few other locals. Thus when newer, perhaps more productive or commercial varieties came along, it is easy to see how human nature might take the approach of 'out with the old and in with the new'. And so trees were felled or, when they died, replaced with different varieties. The people of the time would have thought nothing of replacing one variety with another, there being little room for sentiment in their hard and busy lives.

Does it matter that a number of old fruit types have disappeared? Well, in the greater scheme of things, perhaps not. But all these old varieties are part of our heritage as well as contributing to the diversity of the *Malus* gene pool. As we saw in Chapter 3, around half of the apples grown here today are of that single variety 'Gala', and around half of the remaining ones are 'Braeburn'. Hence three quarters of the apples we eat are of only two varieties, not very healthy from a diversity point of view, especially if a new disease was to arise that ripped through those varieties. So keeping a large diverse gene bank of all our fruit is an insurance policy; should disease devastate the commercial varieties it is that historic gene pool where we would go to find good health, resilience and hardiness.

Whilst many of those historic named varieties are not known today, there are equally still numbers of old and ancient fruit trees hanging on in gardens, forgotten field corners and derelict orchards, which introduces the concept of the orchard or fruit detectives. Those of us that are interested in such things try to identify these remnant trees to see if we can match the evidence to the written records and perhaps rediscover some of these heritage and historic varieties. The credit for this interest has to go to Common Ground who lit the spark that grew into this curiosity of that which is local and distinct. Off the back of their work a number of regional fruit groups were formed that set about attempting to catalogue the varieties of their area and see if any of the lost ones could be rediscovered in those forgotten corners of our gardens and countryside.

For me there is a fascination in finding old trees and seeing if we can identify what they are, helped by the way research has come on in leaps and bounds so that we now have the addition of DNA testing to aid us. All known apple and pear varieties here in the UK have had their DNA profiles logged, along with many across Europe, so any unknown finds can be tested and matched. Thus anything that comes back with a 'no match' points towards either a random seedling or one of those once written about varieties. Accepting that there are no DNA profiles for the missing varieties of the past, it then becomes a desk-based project of looking at old texts, catalogues, pomonas and illustrations to see if an identification can be secured. It can be far from easy, as old, ancient trees often produce very poor quality fruit, undersized and diseased. Poor specimens make identification extremely hard. In some cases we take graft wood from the source tree and grow three or four young versions of those trees which when they crop at three or four years old produce fruit far more representative, or 'to type' as we say. Identification can be a long process, sometimes taking years.

The Marcher Apple Network is the group in my part of the country, founded in 1993 by eight individuals interested in the varieties of the English / Welsh border region and keen to conserve them. Membership has grown to 300 and the group holds three museum orchards containing some of the rarest apples

Through sections, stalk, eye and skin can all be a means of attempting an identification

in the country. Over the years they have attended countless Apple Days, country fairs and shows and put in thousands of hours of research and, in so doing, have rediscovered a number of lost varieties as well as unearthing a range of stories and mysteries surrounding apples of the Marches, as well as those further afield.

The process often goes something along these lines: a member of the public turns up at an Apple Day or country show with an apple from their garden or orchard. If not something the attending group members can easily identify, it is taken away and pored over during an identification workshop using a range of books, keys and comparisons of similar varieties. The apple may be cut through in both planes so the core shape and profile can be studied; the stalk and the eye can offer clues as can the skin and its lenticels (the little pores in the skin). After much deliberation, sometimes a conclusion is drawn and the owner informed. More recently if there is still uncertainty, a DNA sample may be submitted which may or may not give a match. Interestingly the DNA profile is done via leaves taken in mid-summer and not from the fruit. Fruit of the same variety can appear very different when grown in differing soils in different parts of the country, so DNA sometimes solves the matter. In no match scenarios, we may graft and grow on specimens of the young trees in our museum orchards, as open case files, as it were. At some future time, someone else may submit an apple that matches this one's DNA and new light may be shed on the case. An orchard detective's work is never completely done.

David Spilsbury (left) and the author (right) of the Teme Valley Apple Group doing some apple detective work

To illustrate how difficult identification can be, these apples all came off the same tree on the same day; they are the variety 'Blenheim Orange'

Some varieties are easier than others to identify. 'Worcester Pearmain' has these distinctive ball bearing beads around the eye.

I have been lucky enough to get to know a few individuals of a generation, of what we might term the old orchard men and women, those now in their eighties and nineties who worked in orchards in the forties to seventies. All of them were only too pleased to share their knowledge, stories and experiences with me. They are the last of a generation who still recall the era of fruit distribution on steam trains and shire horses working on the farms, of a time when many rural folk were born, grew up and lived in the same tight geographical area all their lives, resulting in an in-depth knowledge of their locality and its history. To drive around the district with one of them who could point at various fields with a "that used to be an orchard and that was an orchard over there" was to appreciate the scale of what has been lost in a generation. Their knowledge has been invaluable in tracking down and rediscovering several of the West Midlands lost varieties. Sadly several of those I got to know have now 'gone on' as they say in these parts but it was a privilege to know them and share a small part of their knowledge.

Between them these regional fruit groups of committed enthusiasts and that older generation have put in the hours and rediscovered a number of varieties once considered lost. However, there is also much that may never be resolved; in the case of pears, the extraordinary statistic is that in 1850 Scott's Nursery of Somerset had a collection of 1500 named varieties of pear, and today the national fruit collection has just over 500, hence around 1000 named pear varieties are lost or unaccounted for. The catalogues of the past give names but little else by way of details, so when we find an old pear tree that does not have a DNA match with anything else, then it may well be one of those lost ones, yet which one, we will likely never be able to ascertain as there simply isn't enough information to work with. Some aspects of fruit history are seemingly forever lost in the mists of time.

INTO THE FUTURE

So much of the subject of orchards is about its deep and fascinating history, wistful stories of bygone eras, that as we have seen were often far from rosy or romantic. But now we are entering a future far from the days of old, an almost science fiction world of DNA and in-depth scientific research. Identifying fruit, particularly apples, has always been problematic. Firstly, there are those thousands of different varieties; frankly no one can know them all. Then add in the fact that an old or diseased tree or trees grown on different soils and in different climatic locations can produce fruit of incredibly diverse appearance, even when they are the same variety. So a foolproof method of identification and classification was always an aspiration among pomologists. DNA is beginning to offer us that system. As with detective work of the criminal nature, getting a DNA fingerprint is definitive and incontrovertible. So the National

Left: The Marcher Apple Network getting busy with identification at a show

Below: 'Witney's Kernel', a lost Worcestershire variety rediscovered by the Teme Valley Apple Group

Witney's Kernel

Orchard detectives, Teme Valley Apple Group members with an ancient 'Haughty's Red' tree, the only one found in its native Worcestershire thus far

The mystery pear Shropshire 1958 and Worcestershire 2019, the only recorded specimens

Fruit Collection of apples, pears and plums have now had their varieties DNA logged, enabling any unknowns to also be tested and cross-referenced against them. This was initially done via leaf samples taken in high summer but is now being developed to be carried out from twig samples, so taking the seasonality out of it. Many fruit identification mysteries can be solved in this way by an exact match, so even if the fruit looked very different, due to soils, age or disease, the DNA does not lie. Equally other submissions come with a 'no match' result meaning that nothing thus far in the DNA log is the same. The beauty of this ever expanding DNA database is that in future, someone else may submit a sample with a match that, even if we don't have a name, may open another avenue of investigation involving links between areas, growers or historic sites.

One such example was a pear tree that I came across on an ancient farm in west Worcestershire. The owner thought it was a 'Worcester Black' pear but to my eye it clearly wasn't so we sent it for a DNA test and the result came back as no match to any named variety. The National Fruit Collection had one, however, because a gentleman from a large Shropshire estate had sent it to them in 1958. Not knowing what it was, they had grafted a couple and kept it in their collection. We have no name and as far as we know those are the only two records of it, Shropshire 1958 and west Worcestershire 2019. It is one of those hard cooking pears popular in Tudor times, but it is clearly not a random seedling and is a variety. Will more evidence one day come to light? Who knows, but the ever expanding world of fruit detection is fascinating even if now somewhat in the realms of sci-fi.

Epilogue

Whether your orchard is large or small, is planted in a single season or added to over a few years, there will come a time when, from a planting point of view, it is complete. From then on, it is a case of what constitutes that annual cycle of orchard work and maintenance. Personally, I find this tree care – the tending and looking after – immensely satisfying. The difference between a well-cared for tree and one planted and left to get on with it can reveal a huge contrast. Aside from the pruning discussed earlier, there are the range of orchard jobs to put on the calendar each year.

There is something amazing about creating something out of nothing; to take an overgrazed corner of a grass field as I did, or a neglected garden, or allotment, and turn it into a bountiful, beautiful and productive space, is up there with the best we can achieve as humans. We live in a world of endless doom and gloom about the state of the environment, and whilst we cannot easily recreate pristine rainforest or clean oceans, we can take those denuded and bland parcels of land and enrich them for human use and biodiversity.

As I stated at the outset, orchards can be wonderful places and there is a magic in creating your own. So when the work is done, it is important to enjoy it, take time to just wander amongst the trees, admire the blossom, look at the insects and marvel at the crop. I never think of working in my orchard as being work because I enjoy it. We have come to see the whole notion of work as something arduous, unpleasant and where possible to be avoided. Yet good work is a wonderful thing, a sense of achievement, of creating and doing that which is constructive and productive. It took me a long time and the coronavirus pandemic to find a time to just sit, to just be, to take time to think, reflect and soak it all in, and in its seventh year, my field orchard seems to be coming of age, almost complete, and aside from that tree care element, able to do its thing.

The bulk of this book was written during the COVID-19 pandemic lockdowns of 2020 and 2021, without doubt the strangest time of all our lives. What it did do, however, was give me the gift of time to write, largely freed of other distractions as well as giving me a focus at that difficult time.

I have been involved with orchards for over 25 years and taught countless workshops and courses on the topic, written countless articles and two previous books* so what I tried to do was pull together all the threads of those experiences, courses, articles, observations and experiences into one place. It was a cathartic, interesting and at times challenging experience to revisit all those different strands and try and weave them into a logical and sequential cloth. I hope to have gone some way to succeeding in that.

What it's all about – my daughter aged three munching a homegrown organic apple

In 2020 and 2021, I spent more hours in my orchard than any other previous year, some of it carrying out those annual tasks, but much of the time just pottering around even sitting on the bench and reflecting, something I had not ever had the time to do before. So for all the benefits of orchards explored in this book, there is also a wonderful wellbeing aspect to having an orchard; nurturing my trees, seeing them grow, enjoying the blossom, the wildlife that comes to the orchard, picking my own fruit, making some juice and simply spending time in the orchard is good for the mind and the soul. My garden and my orchard are jointly my favourite places in the whole world, the places I am most content, most focused and most relaxed. For me having an orchard has been and is so much more than just a few fruit trees growing on a piece of land.

* *The Apples and Orchards of Worcestershire* (2017) and *The Worcester Black Pear* (2018) both by Wade Muggleton available from www.marcherapple.net/store

Glossary of Terms

Permaculture Originated in Australia in the 1970s, the word being an amalgamation of permanence and agriculture. It is essentially about trying to create sustainable ways of living, working with nature and building systems that do not degrade or deplete the earth.

Pomona A book on fruit, most often apples and pears, usually beautifully illustrated with water colours and illustrations – the best examples were produced in the mid to late 1800s.

Apple anatomy:

Base The bottom of the fruit out of which the stalk protrudes, thus the true top of the apple is the eye as that was the flower and thus anatomically the top.

Basin The depression in which the eye is situated, can be deep or shallow, ribbed or smooth and russeted.

Cavity The depression in which the stalk sits.

Eye This is the remaining part of the blossom flower once the fruit is formed, effectively the top of the fruit with the stalk being the bottom.

Lenticels Effectively the pores in the skin, which allow gaseous exchange between the interior and the exterior.

Russet A brownish, grey velvet-like texture that some varieties have a coating of on the skin, to the extent that some carry it in their name e.g. Egremont Russet.

Stalk The means by which the fruit is attached to the tree, can be long or short, and can be an integral feature of identification.

Grafting terminology:

Pippin A rather general term implied to mean a tree raised from a pip as opposed to a grafted tree; when used in a variety name it is supposed to imply the original tree was a random pip rather than an intentionally bred variety.

Rootstock A set of roots usually of the same species onto which the scion is grafted or budded.

Scion A cutting from the parent tree that is then grafted onto either a rootstock or another tree.

Bibliography

As with acknowledgements, as authors we are the living, breathing embodiment of countless things read and absorbed, snippets of information and anecdotes consumed and digested. Some well remembered and referenced but far more often all blended together in a scrapbook-cum-collage of all that which collectively makes up a subject about which we are passionate.

So here is as long a list as I can recall of the books and articles that have helped shape me on my orchard journey.

Arbury, J. and Pinhey, S. (1997) *Pears*, Wells and Winter
Atkins, D. (1992) *The Cuckoo in June*, Thorpe Ltd, Leicestershire
Brown, P. (2016) *The Apple Orchard*, Penguin
Bultitude, J. (1982) *Apples*, Macmillan, London
Bunnyard, E.A. (1920) *A Handbook of Hardy Fruits*, John Murray, London
Clark, M. (2015) *Apples: A Field Guide*, Tewin Orchard, Hertfordshire
Clifford, S. and King, A. (2007) *The Apple Source Book*, Hodder and Stoughton, London
Common Ground (2000) *The Common Ground Book of Orchards*
Crawford, M. (1996) *Directory of Pear Cultivars*, Agroforestry Research Trust
Dunn, N. (2010) *Trees for Your Garden*, The Tree Council, London
Evelyn, J. (1678) *Silva*, London
Flowerdew, B. (1995) *Complete Fruit Book*, Kyle Cathie, London
Hart, R. (1991) *Forest Gardening*, Green Books
Hogg, R. (1884) *The Fruit Manual: A Guide to the fruit and fruit trees of Great Britain*, Journal of Horticulture office
Jacobsen, R. (2014) *Apples of Uncommon Character*, Bloomsbury, New York
Juniper, B. and Mabberley, D. (2006) *The Story of the Apple*, Timber Press, Oregon
Macdonald, B. and Gates, N. (2020) *Orchard: A Year in England's Eden*, William Collins
Mollison, B. and Holmgren, D. (1978) *Permaculture One*, Tagari
Mollison, B. (1979) *Permaculture Two*, Tagari
Mollison, B. (1991) *An Introduction to Permaculture*, Tagari
Morgan, J. and Richards, A. (2002) *The New Book of Apples*, Elbury, London
Morgan, J. (2015) *The Book of Pears*, Elbury Press
Parkinson, J. (1629) *Paradisi in Sole Paradisus Terrestris*
Porter, M. and Gill, M. (2010) *Welsh Marches Pomona*, Marcher Apple Network, Wales
Rea, J. (1676) *Flora, Ceres and Pomona*, George Marriot
Sanders, R. (2010) *The Apple Book*, Frances Lincoln, London
Smith M.W.G, (1971) *The National Apple Register of the United Kingdom*, MAFF, London
Taylor, H.V. (1936) *The Apples of England*, Crosby Lockwood and Sons Ltd.
The Apple and Pear conference report of 1888
Marcher Apple Network (2002) *Apples of the Welsh Marches*

Useful Sources of Information

East of England apples and orchards project www.applesandorchards.org.uk

Fruit ID – helps identify fruit cultivars as part of efforts to conserve heritage varieties www.fruitid.com

Gloucestershire Orchards Trust have done some great work on researching and restoring the county's orchards www.glosorchards.org

People Trust for Endangered Species – have some great work on traditional orchards https://.ptes.org

Permaculture magazine – inspiring articles written by leading experts alongside the readers' own tips and solutions. www.permaculture.co.uk

Permaculture magazine's YouTube channel, including two videos from the author: How to grow fruit trees in small space: www.youtube.com/watch?v=IFf2z8ikXLY&t How to plan and plant your own orchard: www.youtube.com/watch?v=nO1hHMT-tgI

The Marcher Apple Network – have for the last 28 years researched and attempted to rediscover the heritage varieties of the Midlands and Wales, www.marcherapple.net is a rich source of information.

The National Fruit Collection at Brogdale, the treasure house of UK fruit varieties. The collection can be searched through by name and is a great tool to aiding identification. www.nationalfruitcollection.org.uk

The Northern Fruit Group – promoting fruit growing in the north of England www.thenorthernfruitgroup.com

The Orchard Project – originally the London Orchard Project before they went national. Their website has lots of good information www.theorchardproject.org.uk

NURSERIES

www.walcotnursery.co.uk – excellent supplier of organic fruit trees including many rare and heritage varieties.

www.tomtheappleman.co.uk – good source of varieties of the Midlands and Wales.

www.frankpmatthews.com – supplier of wide range of top quality fruit trees as well as other tree species.
www.keepers-nursery.co.uk – claim to have the widest range of fruit tree varieties.

Acknowledgements

Any project or book is an amalgam of countless influences and contributions from all manner of good folk who willingly contribute their knowledge, enthusiasm and good grace. Some influences are obvious and direct, others subtle and almost subconscious. As such there are a raft of people to whom I am indebted for so freely sharing their passion, enthusiasm and stories down the years. One of the joys of this subject is the generosity and enthusiasm of the people who share this passion for orchards and their fruit. To anyone else I have overlooked or forgotten, I apologise.

To my fellow enthusiasts in the Marcher Apple Network who have put in countless hours to rediscover and celebrate the heritage varieties of the region: Ainsleigh, Sheila, Mike and Chris, Andy, David, Daniella, Tom, John, Peter, Jackie, Celia and the late founders. Thank you.

To David Spilsbury, David Powell and the late Reg Farmer and John Edwards who so generously shared their knowledge of the orchards of old and the other members of the small but enthusiastic Teme Valley Apple Group.

To my good friend Andrew M. for his proofreading and his good counsel on all things apple related over many, many years.

To Kevin O'Neil for running the fantastic Walcot Nursery and championing the propagation and selling of local varieties when few others did (www.walcotnursery.co.uk).

To Nick Dunn, Gary Farmer, Tim Dixon and John Edgely from all of whom I learnt various facets of this endlessly fascinating subject.

To Sue Clifford and Angela King, once of Common Ground, without whom this whole interest in the locally distinct and the mapping of local fruit varieties would likely never have come about. Sue and Angela ... yes it really is all your fault.

To Maddy Harland, my editor, for steering this at times erratic ship when it was seriously in danger of going off course.

To Eva Muggleton and Steve Hughes for additional photography.

To my father who made the field project possible by leaving me the funds to buy this little piece of Shropshire and create my orchard. Whilst he obviously never saw it, I hope he would have liked it.

And finally to my family Rache and the kids, Eva and Nils, who probably know more about apples and fruit trees than virtually any other kids their age ... I can only apologise. But hopefully throughout life you will always know a good apple when you see or taste one.

Index

altitude vii, 42, 80
apples v-vi, ix, 1, 3, 10-9, 21-5, 27, 33-5, 39,
    43, 48, 50, 52, 54, 56-7, 60, 66, 69,-72, 74-
    5, 83, 86, 89-90, 98, 101-7, 113, 127, 130-2,
    134-6, 141-3
Apple day viii, 130
apricots 64-5, 130
arch 86-7

bees viii, 49, 120, 124
bullace 60

cherries 63
citrus 66
comfrey 44, 46
Common Ground v, 69, 128, 130, 134, 144
conversion 74
cordon 35, 69
cottage garden 16, 45
currants 44-5, 46

damsons 61, 89, 103, 114, 124
deadwood 120-1
deer 40, 42, 81-2, 97, 120
DNA 12-3, 103, 134-5, 137, 139
dwarf bush 35

espalier 35

figs 65
forest garden 44
forestry tubes 82

grafting 14, 19-20, 42, 72-5, 90
greengages 89
guard (tree) 42, 81-2

hedges 46, 83-4
herbicide 95, 111

livestock 1, 14, 28, 48, 97, 99

Malus 12-3, 16, 19, 42, 48, 71, 74, 83, 131, 134
Marcher Apple Network 133-4, 138, 145
medlar 46, 64, 89
mulberries 66
mulch 28, 44, 46, 52-3, 92, 95-6, 98, 119
mycorrhizal 40, 93-4, 96, 119

nectarines 64

orchard:
    field vi, 38
    garden vi, 32
    inter vi, 42
Orchard Project, The 130, 145

pears 55, 59, 142, 144
permaculture iv, ix-x, 7, 27, 31, 34-6, 38, 40,
    42, 44, 69-70, 77, 128, 145
pitchers 77
plums 4, 11, 13, 28, 44, 52, 60-61, 63, 68-70,
    79, 89, 101, 114, 124, 139
pollination vi, 49
pollinator 25, 123
poultry 35, 37
propagation v, 20, 72
pruning viii, 112-3, 115, 117
prunings 92
Pyrus 13, 51, 74

quinces 51, 63, 89

raspberries 45-6
rootstock 19, 33-4, 39, 47-8, 50-1, 68, 70-4,
    80, 86, 88, 143

scion 39, 48, 72-5, 143
scythe 97, 100
seedlings vii, 71
semi-dwarfing 34-5, 39, 47, 51, 85
sheep x, 1-2, 7, 42, 81-3, 98-100, 111, 127
shelterbelt 83
sloes 60, 103
Station Road Permaculture Garden iv, x,
    32, 38, 46
stepover 45, 85-6
sward 42, 98-9, 124

tools vii, 91

vines 66

wild apples 12-3
wildflowers 36, 123-4
wildings 60
wildlife viii, 1-3, 7, 10-1, 27, 47, 60, 83, 92, 97,
    109, 113, 117, 120, 123-8, 130, 142
wild service tree 67
windfalls 109

Enjoyed this book?
You might also like these
from Permanent Publications

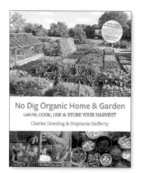